Greening the Firm

Over the last two decades environmental issues have become important in public and business policy. This book asks why firms sometimes voluntarily adopt environmental policies which go beyond legal requirements. It employs a new-institutionalist perspective, and argues that existing explanations, especially from neoclassical economics, concentrate on external factors at the expense of internal dynamics. Prakash argues that "beyond-compliance" policies are due to two types of intra-firm processes, which he describes as power-based and leadership-based. His argument is supported by analysis of ten cases within two firms – Baxter International Inc. and Eli Lilly and Company – including interviews with managers, and access to meetings and documents. This book therefore examines the internal working of firms' environmental policy in a theoretically rigorous way, providing a significant contribution to the theory of the firm. It will be valuable for students of business and environmental studies, as well as political economy and public policy.

ASEEM PRAKASH is Assistant Professor, Department of Stategic Management and Public Policy at the School of Business and Public Management, Department of Political Science, and The Elliott School of International Affairs, The George Washington University.

Greening the Firm

The Politics of Corporate Environmentalism

Aseem Prakash

The George Washington University

PUBLISHED BY THE PRESS SYNDICATE OF THE UNIVERSITY OF CAMBRIDGE
The Pitt Building, Trumpington Street, Cambridge CB2 1RP, United Kingdom

CAMBRIDGE UNIVERSITY PRESS
The Edinburgh Building, Cambridge CB2 2RU, UK http://www.cup.cam.ac.uk
40 West 20th Street, New York, NY 10011-4211, USA http://www.cup.org
10 Stamford Road, Oakleigh, Melbourne 3166, Australia

First published 2000

Printed in the United Kingdom at the University Press, Cambridge

Typeset in Plantin 10/12 pt in QuarkXPress™ [SE]

A catalogue record for this book is available from the British Library

Library of Congress Cataloguing in Publication data

Prakash, Aseem.
Greening the firm: the politics of corporate environmentalism / Aseem Prakash.
p. cm.
Includes bibliographical references and index.
ISBN 0 521 66249 4 (hardbound)
1. Environmental policy Case studies. 2. Industrial management – Environmental aspects Case studies. 3. International business enterprises – Environmental aspects Case studies. 4. Baxter International Inc. 5. Eli Lilly and Company. I. Title.
GE170.P69 2000
658.4'08–dc21 99-33723 CIP

ISBN 0 521 66249 4 hardback
ISBN 0 521 66487 X paperback

To Papa and Mummy

Contents

List of tables

List of appendixes

Preface

This project has a long history. After completing my MBA from the Indian Institute of Management, Ahmedabad, I joined Procter and Gamble's marketing department in 1989. I was a great believer in the dominant paradigm that is often taught at business schools: firms maximize profits, those that do not are punished by the market, and managers can objectively pursue profit maximization by employing tools of investment appraisal. So, armed with technocratic knowledge and naive enthusiasm, I joined Procter and Gamble that was (and continues to be) a well-regarded and profitable company. One of the things that really struck me was the role of organizational dynamics in affecting a firm's tactical and strategic decisions. The neoclassical economic theory and the various sophisticated financial and marketing techniques that I had learnt at my alma mater did not seem to have the desired relevance. I had always thought that only the functioning of governmental bureaucracies was impacted by organizational politics. How wrong I was! As I exchanged notes with my fellow MBAs across firms and functional areas, I realized that they too were having similar experiences. Broad structural factors external to firms indeed outlined a framework within which firms made decisions. Firms also seemed to pursue a loosely defined objective of "high" levels of "profits." However, internal politics – inter-personal, inter-departmental, etc. – was important in shaping outcomes. Many projects that were pursued were clearly wrong and many "sensible" policies were not adopted. The strategies and power of key individuals mattered in shaping organizational outcomes.

By 1992, I was ready for a change. Also, Anil Gupta, Indian Institute of Management, Ahmedabad, whom I admire very much, had been encouraging me to pursue a career in academia. In 1993, I enrolled in the Joint Ph.D. program co-sponsored by the School of Public and Environmental Affairs and the Department of Political Science at Indiana University, Bloomington. I planned to focus on environmental issues and it seemed that I had an opportunity to leverage my experience of the corporate world for this task. I, therefore, decided to examine how firms actually make environmental policies. Since most large firms in the US are voluntarily adopting programs that are more stringent than the requirements of the extant laws, I decided to focus on firm-level response to "beyond-compliance" policies. Thus, my research project seemed to have important business strategy and public policy implications.

Beyond-compliance initiatives are also interesting to study because they provide a political space for discursive struggles among managers on the benefits and costs of various environmental programs. As a consequence of such struggles, profits no longer remain an objective and abstract concept that can be technocratically determined. They become a subjective concept that are influenced by varying and changing managerial interpretations of future benefits and costs. As I discovered during my field work, many beyond-compliance projects, even the ones costing hundreds of millions of dollars, were not subjected to formal investment appraisal procedures. Organizational dynamics played an important role in deciding whether or not such projects were adopted. I could discern various kinds of organizational dynamics, the conditions under which they were activated, and their impact on the policymaking processes and on the final policy outcomes. Thus, "unpacking" firms to study environmental policymaking processes provided interesting theoretical and policy-relevant insights.

In summary, the contributions of this study are fourfold. First, it highlights the inadequacy of the neoclassical theory, the dominant paradigm, in explaining why firms selectively adopt beyond-compliance policies. Second, at a broad level, it argues that "agents" (or managers) have some (not complete) autonomy in pursuing beyond-compliance policies; external "structures" alone cannot provide fully specified explanations. Organizational dynamics play an important role. Third, it focuses on the important role of power-based and leadership-based processes in shaping the policies of firms. It argues for "bringing back leadership" in the study of political economy. Further, the book integrates insights from sociological institutional theory and stakeholder theory (that focus on pressures external to firms) with leadership-based and power-based theories of the new-institutionalist tradition. Finally, since the conclusions of this book are generalizable to other issue areas where firms adopt beyond-compliance policies, it outlines important questions for future research.

At Indiana University, I had the privilege of working with the leading scholars of environmental policy, institutional analysis, and political economy. Elinor Ostrom, my mentor, encouraged me to embark on this multi-disciplinary project. As the dissertation committee co-chair and the first reader, she constructively challenged me throughout the process. Lin also provided me with valuable office space at the Workshop in Political Theory and Policy Analysis that enabled me to work in the Workshop's intellectually rich environment and interact with scholars from all over the world. Jeffrey Hart, also my committee co-chair, gave me very useful input on the dissertation and then continually encouraged me to revise and refine it into a book manuscript. Mike McGinnis contributed significantly in shaping this project as well as influencing my overall intellectual agenda. Rosemary O'Leary supported this project from the very outset and introduced me to the idea of examining environmental policies of Baxter International and Eli Lilly. Kerry Krutilla helped me to understand the political economy of environmental issues. Vincent Ostrom has been a source of inspiration and continued intellectual support.

Internal workings of firms on environmental policies are difficult to examine because firms are often unwilling to share information on this subject. I am indebted to Dean James Barnes (School of Public and Environmental Affairs) who wrote to Baxter and Lilly on my behalf, thus paving the way for my empirical work. Managers at Baxter and Lilly were first intrigued by the objectives of my research. This was not surprising since this was the first occasion that an academic researcher was closely scrutinizing them. Eventually, they were very supportive of the project. They candidly discussed policy dynamics of various environmental programs. They made themselves available for multiple interviews and follow-up questions. I had opportunities to attend their environmental policy meetings and to examine their internal documents. However, the interpretations and analyses of the events pertaining to environmental policymaking are entirely mine. Bill Blackburn (Baxter) and Daniel Carmichael (Lilly) enabled me to "step inside" their respective organizations. Don Brannon (Lilly) was extremely helpful in ensuring access to appropriate managers as well as documents. Bill Blackburn, Ron Meissen (Baxter) Don Brannon, and Joan Heinz (Lilly) read my dissertation and provided useful input. I would also like to acknowledge help and support from the following: Rob Currie, Mike Cycyota, Peter Etienne, Munira Meharauli, and Verie Sandborg (all from Baxter); Jonathan Babcock, Catherine Ehlhardt, Bert Gorman, Rick Lattimer, Ronald Pitzer, James Vangeloff, and John Wilkins (all from Lilly).

This book is a significantly revised version of my doctoral dissertation. John Haslam, Cambridge University Press, was very supportive and helpful throughout the review process. The two anonymous referees provided excellent and constructive comments that brought greater clarity to my arguments. My friends and colleagues at Indiana University and George Washington University patiently read and reread numerous drafts and gave me invaluable substantive and editorial comments. These are:

Arun Agrawal, Trevor Brown, Brenda Bushouse, George Candler, Yu-che Chen, Nives Dolsak, Ray Eliason, Dan Kane, Tom Koontz, Matthew Potoski, Susan Sell, and Cynthia Yaudes. Jun-ho Choi and Jennifer Baka provided valuable research and editorial support.

This project involved making numerous visits to Indianapolis (Lilly) and Chicago (Baxter). To defray the costs of fieldwork and other expenses, I received funding from the Center for International Business Education and Research, Kelley School of Business, Indiana University, and the College of Arts and Sciences, Indiana University.

Finally, my family supported me emotionally, intellectually, and financially. I have been fortunate to have wonderful parents who have supported all my ventures including the decision to abandon the lucrative world of Procter and Gamble for a career in academia. I dedicate this book to them. My brother, Anshu, and my sister, Charu, have been a source of great support and affection. They have permitted me to enjoy all the benefits of being the youngest in the family. Nives Dolŝak stood by me during my stay at Indiana University and also conceded to do so in the future by marrying me.

1 Greening the firm: an introduction

Though environmental problems have challenged humankind since time immemorial, policy scientists have given serious attention to environmental issues only since the 1960s. A series of industrial accidents and media events such as the publication of Rachel Carson's *Silent Spring* (1962) highlighted the environmental consequences of unfettered industrialization. Responding to public concerns, from the 1970s onwards, the United States Congress has enacted a series of laws stipulating environmental standards and technologies for firms. These policies were often backed by zealous monitoring and enforcement. In the 1980s the policy community and the regulatees began articulating their dissatisfaction with the inefficiencies of command and control policies, specifically questioning the capacities of governmental agencies to implement detailed regulations.

Since the late 1980s, particularly after the Rio Summit of 1992, policymakers appear to have accepted that governmental coercion alone will not be sufficient in forcing firms to adopt environmentally sustainable policies; "right incentives" must be provided (Hahn and Nell 1982; Lee and Misiolek 1986; Baumol and Oates 1988; Oates, Portney, and McGartland 1989; Atkinson and Tietenberg 1991; Tietenberg 1992). More recently, policymakers are beginning to play down their adversarial role, and are highlighting the potential gains of collaborating with firms in developing and implementing environmental policies. Further, as opposed to a reluctance in implementing environmental laws, firms are increasingly inclined to adopt "beyond-compliance" environmental policies, the ones that are more stringent than the requirements of the extant laws and regulations.

Beyond-compliance initiatives could be designed and implemented by regulators, industry associations, or individual firms. For example, in recent years, the United States Environmental Protection Agency (EPA) has launched voluntary beyond-compliance programs such as Green Lights, Project XL, and 33/50. These initiatives are win–win–win for the regulators, firms, and citizens. Regulators are able to implement their mandates to enforce environmental laws at low costs. This is particularly welcome in an era of declining budgets for many governmental programs and of calling for "reinventing government" (Osborne and Gaebler 1992). Citizens enjoy cleaner air and purer water without an increased tax burden. Firms enjoy greater operational flexibility in designing and implementing environmental programs that the command and control era denied to them. Their relationship with regulators also becomes less adversarial.

Regardless of whether regulators view firms as adversaries or as potential partners, as reluctant implementers of extant laws or as enthusiastic participants in beyond-compliance programs, environmental policy scientists have implicitly treated firms as unitary actors with similar responses to external incentives (notable exceptions include Fischer and Schot 1992; Gable 1994; Bunge, Cohen-Rosenthal, and Ruiz-Quintanilla 1995). As a result, there is an inadequate understanding of the internal processes that lead firms to adopt or not adopt various kinds of environmental policies, especially the beyond-compliance ones. In other policy areas and disciplines, however, firms have been "unpacked" and their internal processes extensively studied (March and Simon 1958; Baumol 1959; Cyert and March 1963; Marris 1964; Williamson 1964; Katz and Kahn 1966; Thompson 1973; also, Allison 1971). There is also a well-established literature on the impact of external factors on intra-firm dynamics (Cyert and March 1963; Pfeffer and Salanick 1978; DiMaggio and Powell 1983; Tolbert 1985; Oliver 1991).

In contrast to the existing environmental policy literature, this book examines the processes of environmental policymaking within firms. The *theoretical question* I address is: why do firms *selectively* adopt beyond-compliance environmental policies? Selective adoption implies that a given firm adopts only some but not all policies with similar characteristics, or different firms within the same industry respond differently to a given policy. This study argues that the existing explanations that focus exclusively on factors external to firms and that treat firms as unitary actors are under-specified to answer this question. An examination of intra-firm dynamics is also required. Though factors external to firms create incentives and expectations for managers, intra-firm politics influences how managers perceive and interpret external pressures and act upon them. My *policy question* therefore is: why and how do external factors aid or thwart supporters of beyond-compliance policies to persuade their firms to adopt these policies?

To examine these questions, I explore the following issues. How do managers make decisions on environmental policies? What are the decision criteria? Do managers have different preferences on environmental policies and, if so, do such differences impact policy adoption? Are beyond-compliance policies adopted only if they are projected to deliver adequate levels of quantifiable profits? How are non-quantifiable benefits brought into the equation? Since answers to these questions vary within and across firms, the book investigates internal processes and inter-managerial interactions on environmental policymaking.

Beyond-compliance: an overview

Beyond-compliance is different from over-compliance. In the latter, firms seek to comply with the law but due to technological indivisibilities, deliver more than the legal requirement. Also, adopting uniform technologies across facilities that face varying environmental regulations results in over-compliance (Oates, Portney, and McGartland 1989). In contrast, beyond-compliance policies specifically propose to exceed the requirements of extant laws. They may involve modifying physical aspects of value-addition processes or adopting new management systems.

The profit-maximizing view of the firm predicts that firms will adopt policies that can be demonstrated, *ex ante*, to meet or exceed firms' profit criteria. Thus, from a managerial perspective, environmental policies can be classified along two attributes: (1) whether they meet or exceed the *ex ante* profit criteria as stipulated in capital budgeting or some other established investment appraisal procedure; (2) whether they are required by law or they are beyond-compliance. Based on these attributes, four modal policy types can be identified: Type 1 (beyond-compliance and meet or exceed the profit criteria), Type 2 (beyond-compliance but cannot or do not meet the profit criteria), Type 3 (required by law and meet or exceed the profit criteria) and Type 4 (required by law but cannot or do not meet the profit criteria). This discussion is summarized in table 1.1.

Since Type 3 and Type 4 policies are required by law, firms are expected to adopt them. This is especially true for industrialized countries where environmental laws are perceived by managers as being strictly enforced and penalties for non-compliance are significant. Consequently, most firms are not expected to systematically violate environmental laws. This book, therefore, does not focus on these policies.

Type 1 policies, though not required by law, are consistent with the profit-maximizing model of a firm since they meet the *ex ante* profit criteria. For example, scholars suggest that firms can increase profits by voluntarily reducing pollution (Porter 1991; Porter and van der Linde 1995; Shrivastava 1995; Hart 1995; Russo and Fouts 1997; for a critique, see Walley and Whitehead 1994; Newton and Harte 1996). Such policies enable firms to capture the "low-hanging fruit." It is also suggested that

Table 1.1. *Categories of environmental policies*

	Impact on Compliance	
Impact on quantifiable profits	Ensure compliance	Result in beyond-compliance
Established procedures to assess profitability are employed and the policy meets or exceeds their criteria	Type 4 profitable policies that are required by law; are implemented with low inter-manager conflict	Type 1 policies that involve profitable organizational changes with low inter-manager conflicts
Either established procedures to assess profitability cannot be employed, or if they can be, then they were not employed	Type 3 policies that are required by law; are implemented with low inter-manager conflict if there is stringent punishment for non-compliance and effective monitoring	Type 2 policies that involve inter-manager conflicts

such policies enable firms with greater consumer contact to compete on environmental quality and charge a premium (Arora and Cason 1996). Based on these arguments, these policies seem win–win for virtually every constituent. Of course, due to inertia or lack of knowledge about profit opportunities, firms may be slow to adopt them. Nevertheless, serious opposition within firms to such policies is not expected, and, consequently, this book does not examine them.

In contrast to Type 1, 3, and 4 policies, managers are expected to differ on the economic usefulness of Type 2 policies. This book, therefore, exclusively focuses on these policies. Literature identifies multiple motivations for firms to adopt Type 2 policies. The first category of explanations identifies strategic reasons geared towards potential long-term economic benefits. Firms could preempt and/or shape environmental regulations if they themselves adopt such policies (Fri 1992; Khanna and Damon 1999) and reap first-mover advantages (Nehrt 1998; Porter and van der Linde 1995; for a critique, see Palmer, Oates, and Portney 1995; Rugman and Verbeke 2000). Similarly, technologically advanced firms could raise the cost of entry for their rivals – the assumption being that higher standards will lead to stringent regulations (Barrett 1991; Maloney and McCormick 1982; Salop and Scheffman 1983).

The second set of explanations – sociological institutional theory and stakeholder theory – focus on non-profit objectives of firms that may or may not impact their long-term profit objectives. The institutional theory focuses on the impact of external institutions on the policies of firms

(Scott 1987; Zucker 1988; Oliver 1991; Meyer and Scott, 1992; Hoffman 1997). In contrast to neoclassical economics that privileges two institutions – markets and governments – institutional theory takes into account other social institutions as well. Questioning the atomistic accounts of organizational policymaking, it suggests that firms are not profit maximizers; their policies reflect external pressures for legitimacy. Of course, different institutions have varying capacities to influence firms. This theory would predict that firms adopt Type 2 policies in response to pressures from key external institutions and *managers would have little autonomy in this regard* (Hoffman 1997: 6).[1]

Neoclassical economics views the social objective of business is to maximize shareholders' wealth (Friedman 1970). In contrast, stakeholder theory suggests that firms should (and sometimes do) design policies taking into account the preferences of multiple stakeholders – stakeholders being "any group or individual who can affect or is affected by the achievement of the organization's objectives" (Freeman 1984: 46; Donaldson and Preston 1995; Clarkson 1995). Similarly, the literature on corporate social performance (CSP), responsibility, and responsiveness argues that firms have societal responsibilities other than the pursuit of shareholder wealth maximization (Preston 1975). CSP policies are adopted because they are the "right things to do." Firms could be reactive, defensive, accommodative, and proactive in dealing with them (Wartick and Cochran 1985; Carroll 1995; for a critique see, Wood 1991). It could be argued that since Type 2 policies represent proactive CSP, they are adopted by firms.[2] Of course, different stakeholders and institutions have different expectations; sometimes expectations may even be in conflict (Wood and Jones 1995). Thus, it is critical to examine how managers interpret these expectations and employ them to push their agendas on Type 2 policies.

Though institutional theory and stakeholder theory correctly identify non-profit and long-term (potential) profit reasons for adopting Type 2 policies, they inadequately explain *variations* in response – why do firms *selectively* adopt them? For example, why does firm X consider Policy A

[1] Oliver (1991) acknowledges that "agents" may have autonomy even in an institutionalist perspective.

[2] Scholars have examined whether CSP policies positively impact firms' financial performance (Ackerman 1975; Preston and Post 1975; Preston and Sapienza 1990; Jones 1995) and adopting a stakeholder approach furthers firms' economic performance (Cochran and Wood 1984; Barton, Hill, and Sundaram 1989; Kotter and Heskett 1992). These studies have been criticized on theoretical and methodological grounds. As a result, these literatures are inconclusive on the impact of adoption of CSP policies and/or stakeholder approach on firms' economic performance (Wood and Jones 1995; Griffin and Mahon 1997).

but not Policy B as "the right thing to do" although both policies have similar characteristics? Or, why does firm X but not firm Y believe that adopting policy A is the "right thing to do"?

This book draws insights from institutional theory and stakeholder theory and relates them to dynamics within firms. The point of departure is that I do not view managers as passive recipients of external pressures. Since "agents" have autonomy in the realm of Type 2 policies, explanations focusing on external "structures" only are under-specified (Child 1972; Granovetter 1985; Ostrom 1990). Further, managers do not have homogeneous preferences on Type 2 policies. The book focuses on the role of key managers in generating consensus or, if faced with opposition, lobbying the top management to mandate policy adoption. While not denying the importance of external factors, I highlight that in the context of Type 2 policies, managers have autonomy to interpret the impact of external pressures on the long term profit and non-profit objectives. Hence, intra-firm politics is important in explaining variations in adoption within and across firms.

"Unpacking" the firm

To understand internal policy processes, an explication of the notion of a firm is imperative. Neoclassical economic theory treats firms as unitary actors seeking to maximize profits (Hirshleifer 1988). The book interprets its broad message as that firms adopt only those policies and projects that can be demonstrated *ex ante* as potentially profitable. Project appraisal is a technical process and there is a shared understanding among managers about the legitimacy of established appraisal procedures, particularly capital budgeting. This procedure requires estimating future benefits and costs and discounting them with an appropriate discount rate. If a project meets or exceeds a given rate of return, it is deemed potentially profitable. Consequently, capital budgeting ensures that managers examine investment decisions objectively with a focus on maximizing shareholders' wealth. This is an important safeguard for shareholders who often have little say in the running of firms, and are therefore vulnerable to "agency abuses" by managers (Berle and Means 1932; Fama 1980). Further, since maximizing a firm's measurable profits is the primary objective for all managers, policy processes would be consensual. This is not to say that managers have identical preferences on environmental policies. Most likely they do not. However, different managerial preferences are not predicted to play out in the policymaking process because there is consensus that a policy should meet or exceed the profitability criteria.

Capital budgeting is appropriate to assess profitability of projects that involve up-front capital expenditures and generate future cash flows. Since Type 2 policies may not involve capital expenditures, capital budgeting is inappropriate for assessing their profitability. In general, it is difficult to assess the impact on profits of policies that focus on establishing management systems, and hence do not generate revenue or decrease quantifiable costs. To assess the profit impact of Type 2 policies, managers employ subjective methods. Such projects are justified by some managers by arguments such as "they are good for the firm in the long run" and "they are important for keeping the EPA in good humor."

Further, some policies involving significant up-front capital expenditures may be adopted without being subjected to capital budgeting. This suggests that established procedures are not applied consistently and policymaking within firms involves a complex mix of factors. Intra-firm processes, inter-manager interactions, and managerial perceptions of external factors are important in influencing whether or not a Type 2 policy is adopted. Project appraisal is not a technical process only; organizational politics also plays an important role in influencing managerial perceptions of the desirability of a project.

The neoclassical notion of a firm is useful in predicting market outcomes in highly competitive markets or when policies are required by laws that are strictly enforced. It is not helpful in explaining why firms selectively adopt Type 2 policies. For this we need to examine the internal processes of firms. Treating firms as composites consisting of many managers, this book employs a new-institutionalist perspective. Further, it assumes that while maximizing quantifiable profits is often the preeminent goal of most managers, it is not the only goal. Managers also differ in their subjective assessments of the long-term profit impacts of policies. I classify managers into two categories: (1) policy supporters favoring the adoption of beyond-compliance policies whose profit impact is not quantified; and (2) policy skeptics who oppose such policies. There is a third category as well: policy neutral. Since they do not significantly impact policy dynamics, the book does not focus on them.

Within a new-institutionalist perspective, three broad theories of firms can be identified: transaction cost, power-based, and leadership-based. Transaction cost theorists examine an important question that is not adequately addressed by neoclassical economics: why do firms arise at all; alternatively, why and how do managers arrive at "make or buy" decisions? Following Coase (1937), transaction cost theorists view firms as institutions designed to economize on transaction costs by allocating resources through hierarchical fiats and not market mechanisms. Williamson (1975, 1985) focuses on how firms evaluate "make or buy" decisions, and suggests that these decisions reflect managers' desire to minimize transaction costs given asset specificity, bounded rationality, and a potential for labors' opportunism. Transaction cost theories, however, do not specifically address my research question: why do firms selectively adopt beyond-compliance policies? Consequently, the book employs power-based and leadership-based explanations only to examine intra-firm dynamics. Importantly, both theories focus on the crucial role of organizational politics – especially the preferences, strategies, and endowments of key managers – in shaping policy outcomes.

There is extensive literature suggesting that managers are "boundedly rational," often have heterogeneous preferences, and are organized as coalitions that seek different policy objectives (Cyert and March 1963; Simon 1957). Since boundedly rational managers make decisions under uncertainty, decision making is often influenced by inter-managerial interactions. Employing these insights, this book suggests that beyond-compliance policies provide political space for "discursive struggles" (Hajer 1995) within firms on their long-term profit and non-profit impact. If such policies are adopted, it is by two kinds of processes: (1) power based, where policy supporters, in face of opposition from policy skeptics, "capture" the top management and have it mandate the adoption of such policies; (2) leadership based, where policy supporters succeed in inducing consensus, convincing policy skeptics and policy neutrals of the long-term benefits of such policies. It is important to differentiate power-based from leadership-based processes since they arise under different conditions and lead to different types of outcomes. In both processes, managers invoke the external environment in different ways to advocate their policy preferences. The final outcome depends on factors such as policy supporters' hierarchical position, their persuasive or canvassing abilities, their expertise in the issue area, and how they invoke external factors to shape perceptions of others. Policy outcomes would also be influenced by the degree of organizational change required for their implementation: the greater are the predicted changes, the stronger are the incentives for the "losers" to oppose policy adoption. Consequently, the likelihood of policy adoption decreases.

In examining beyond-compliance policies, the book first employs the neoclassical theory: can the profitability of a beyond-compliance policy be assessed by employing capital budgeting? If this theory does not hold (that is, capital budgeting was either inapplicable, or, if applicable, it was not employed), then I turn to power-based or leadership-based theories. Policy processes marked by imposition are classified as power based, and the ones marked by induced consensus as leadership based. The key actors, policy supporters and policy skeptics, are identified and their positions in the hierarchy, and their strategies and logics for supporting or opposing a policy are examined. Since preferences are inferred from behaviors, the book does not examine why policy supporters or policy skeptics have certain preferences. However, it seeks to understand whether policy adoption requires significant levels of organizational changes that upset the status quo, thereby creating incentives for "losers" to oppose a policy. It also examines how factors external to firms support or impede the efforts of policy supporters.

In summary, the theoretical contributions of this study are fourfold. First, it highlights the inadequacy of the neoclassical theory in explaining why firms selectively adopt Type 2 policies. Second, at a broad level, it argues that "agents" have some (not complete) autonomy in pursuing beyond-compliance policies; external "structures" alone cannot provide fully specified explanations. Third, it focuses on the important role of power-based and leadership-based processes in shaping the policies of firms. It argues for "bringing back leadership" in the study of political economy. Further, the book integrates insights from sociological institutional theory and stakeholder theory (that focus on pressures external to firms) with leadership-based and power-based theories. Finally, since the conclusions of this book are generalizable to other issue areas where firms adopt Type 2 policies (often subsumed under social policies), it outlines important questions for future research.

Research designs and methods

At an empirical level, I focus on two firms – Baxter International Inc. and Eli Lilly and Company – and study their key environmental programs during 1975 to mid 1996. Both Baxter and Lilly are multinational enterprises (MNEs). Since MNEs are important economic actors, they have critical roles in environmental policymaking and implementation (Walters 1973; Pearson 1985; World Commission on Environment and Development 1987; Leonard 1988; World Bank 1992; Schmidheiny 1992; Choucri 1993; Jaffe, Peterson, Portney, and Stavins 1995; Prakash, Krutilla, and Karamanos 1996). Therefore, one objective of many environmental policies is influencing the environmental performance of MNEs. This requires an understanding of how MNEs make environmental policies. Unfortunately, there is little literature on this subject as most environmental policy scholars treat MNEs (or any firm for that matter) as unitary actors.

This study focuses on environmental policymaking in the US operations of Baxter and Lilly. It does not study environmental policymaking in their subsidiaries outside the US. For most MNEs operating in industrialized countries, compliance with domestic environmental regulations is often a non-issue though previously many have resisted complying with laws. I attribute such compliance to stringent laws specifying significant civil and criminal liabilities, relatively serious implementation of environmental laws by regulatory bodies and the courts, active monitoring by environmental groups and local communities, and pressures from employees to "go green." The battle within most firms is now being fought in a different arena: to what extent, if at all, should firms go beyond minimum regulations?

Why Baxter and Lilly? According to the largest ever survey of MNEs' environmental programs, these policies are significantly influenced by MNEs' line of business, sales volume, and home country (UNCTAD 1993). These factors are briefly discussed below.

The line of business The high-risk industries as well as the "sun-rise" industries have the strongest environmental programs. High-risk industries such as oil and chemicals have extensive environmental programs because a single industrial accident can inflict significant costs on them. Since sun-rise industries such as electronics, biotechnology, and specialty chemicals have quick product obsolescence, they replace their capital equipment in short cycles. Consequently, they are afforded opportunities to install state-of-the-art, resource-efficient technologies. Further, their high profitability provides them with resources for investing in environmental programs that often have long gestation lags.

The size of MNEs Large MNEs (sales of $4.9 billion and above) have more comprehensive environmental programs than the smaller MNEs because they can tap economies of scale on such expenditures.

The home country of the MNE The scope and content of environmental practices vary significantly across regions. The UNCTAD survey notes that:

> [P]robably the nature of the regulatory environments in the home country of the corporation explains variations. . . . The tendency of Asian corporations [that is, Japanese] to view EH&S [Environmental Health and Safety] activities as business opportunities could be related to the fact that Japanese EH&S policy is formulated to a large extent by the Ministry for International Trade and Industry and not the Environmental Agency. The relatively low utilization of EH&S policies and practices in Europe is probably related to the fact that European environmental regulations tend to rest on administrative enforcement and cooperation between industries. On the other hand, United States' environmental regulation has traditionally been described as adversarial and aggressive, and seems to have encouraged the TNCs [transnational corporations] to establish EH&S procedures to minimize liabilities. (1993: 93)

I have controlled for the three factors identified by the UNCTAD report. Baxter and Lilly share the following characteristics:

(1) Their annual sales exceed $4.9 billion ($9.3 billion for Baxter and $6.7 billion for Lilly in 1995).[3]
(2) They are in the health-care industry.
(3) The United States is their home country.

In addition, these firms are:

(4) significantly globalized (in 1995, the non-US operations accounted for 42.5 percent of Lilly's sales and 28.2 percent of Baxter's sales); and
(5) formally committed to adopting beyond-compliance environmental policies (Baxter 1994a; Eli Lilly 1994a).

At an empirical level, this book examines ten cases of Type 2 policymaking: four common to both firms (underground tanks, 33/50, ISO 14000, and environmental audits), and one each idiosyncratic to them (Responsible Care to Lilly and green products to Baxter).[4] The cases that are briefly described below pertain to policymaking during 1975 to mid 1996.

Underground storage tanks

Underground storage tanks can contaminate soil and ground water creating significant clean-up costs. Consequent to the EPA's regulation in 1985, both Baxter and Lilly removed their existing single-walled underground tanks and installed new tanks that have significant beyond-compliance features. I examine why these firms invested huge amounts of money in beyond-compliance features: about $10 million for Baxter and $30–40 million for Lilly.

Toxic Release Inventory and EPA's 33/50 program

Both Baxter and Lilly took significant beyond-compliance initiatives to reduce their releases of chemicals listed under the Toxic Release Inventory program (TRI). Lilly has invested about $80 million for reducing its releases of TRI chemicals and Baxter has invested about $10

[3] On October 1, 1996, Baxter announced that it had reorganized itself into two corporations: Baxter International Inc. and Allegiance Corporation. Baxter International Inc. focuses on developing medical technologies and Allegiance Corporation focuses on supplying medical and laboratory products (Baxter 1997). Since I am studying environmental policymaking in Baxter and Lilly during 1975–mid 1996, Baxter's reorganization does not affect my research design or analysis.

[4] In chapter 5, external and internal environmental audits as well as Phase I and II of Responsible Care are examined separately.

million for reducing its releases of TRI chemicals, air toxics, and chlorofluorocarbons.

33/50 is a voluntary beyond-compliance program launched by the EPA in 1991. Firms are encouraged to commit to reducing aggregate releases of seventeen specific TRI chemicals, 33 percent by 1992 and 50 percent by 1995 with 1988 as the baseline. Both Baxter and Lilly are charter members of this program and both have exceeded their 1995 objectives.

Chemical manufacturers association's responsible care

The US Chemical Manufacturers Association (CMA) launched Responsible Care in 1988. Under this program, the CMA's member firms are asked to adopt a series of beyond-compliance policies. This case focuses on Lilly only. After initial hesitation on some aspects of this program, Lilly adopted Responsible Care and now is a show-case example of its successful implementation.

"Green" products

Both Lilly and Baxter have adopted a variety of beyond-compliance policies to "green" their manufacturing operations and management systems. However, only Baxter markets green products, the ones that explicitly promise environmental protection as one of their benefits. Given the nature of Lilly's business of manufacturing and marketing ethical or prescription drugs, green products have little business rationale.

Environmental audits

Though there is major controversy over granting attorney–client privilege to environmental audits, both Baxter and Lilly have established strong internal audit programs. In addition, Baxter invites external auditors to evaluate its environmental programs. In 1991, Arthur D. Little was invited to help in defining the state-of-the-art environmental standards and in evaluating whether Baxter's environmental program met those standards.

International Organization for Standardization's ISO 14000

The ISO 14000 series specifies beyond-compliance management systems. These standards have been sponsored by the International Organization for Standardization, a Geneva-based non-governmental organization. ISO 14000 could be viewed as an industrial code of practice that needs to be certified by external auditors. This certification is done at facility level. Currently, such certification is estimated to cost about $20,000 per facility. Neither Baxter nor Lilly have mandated that their facilities should have the ISO 14000 certification; they have adopted a wait-and-see policy. This could be attributed to their extant investments in other industrial codes (Responsible Care for Lilly and the state-of-the-art program for Baxter), and meager perceived gains from switching over to ISO 14000.

Information on these cases was gathered from the following sources: interviews with managers (both in-service and retired), attendance as an observer in meetings of various environmental teams, review of published as well as unpublished documents, and professional journals. Most managers in these firms have been extremely cooperative in sharing information and have not attempted to influence my interpretation of events. However, to maintain confidentiality of my sources, this book does not identify them in any manner, except when quoting from a published document.

Case selection

In examining the above cases, I define the dependent variable as the adoption or non-adoption of Type 2 policies, and the independent variables as factors internal and external to firms. The internal factors include: whether a policy required up-front capital expenditure or whether it involved establishing management systems; the level of expenditures; and the degree of organizational change required to implement a policy. Some of these policies involved significant capital expenditures (underground tanks and 33/50, in particular) and could therefore have been subjected to capital budgeting. Other policies involved establishing management systems (Responsible Care and ISO 14000) whose financial impact cannot be quantified. In addition, the degree of organizational change required for implementing such policies also varied: "minimal" for underground tanks and "significant" for external audits.[5]

To understand the roles of external factors, the book focuses on the managerial perceptions (and how they were shaped) of the abilities of such organizations to impose excludable costs or provide excludable

[5] Minimal changes involve only operational-choice level while significant changes involve collective- and constitutional-choice levels. I discuss the three levels of institutional analysis (operational-, collective-, and constitutional-choice) in chapter 5.

Table 1.2. *Cases and their dimensions*

	External factors encouraging or discouraging this policy	Time period	Scope
Underground Tanks	Encouraged by Environmental Protection Agency	Mid 1980s–late 1980s	Firm
TRI and 33/50	Encouraged by the Environmental Protection Agency	Late 1980s–early 1990s	Manufacturing industry
Responsible Care	Encouraged by Chemical Manufacturers Association	Late 1980s	Chemical industry
"Green Products"	Encouraged by health-care providers, such as hospitals	Early 1990s	Firm
Environmental audits	Discouraged by the Environmental Protection Agency; encouraged by state environmental agencies	Early 1990s–mid 1990s	Firm
ISO 14000	Encouraged by a Geneva-based non-governmental organization	Mid 1990s	All Industries

benefits to firms and to individual managers. The following external factors are examined: governmental agencies (the EPA), non-governmental international organizations (the International Organization for Standardization); industry-level associations (the CMA), and customers (hospitals).

As shown in table 1.2 above, these cases also pertain to different time periods: from the mid 1980s (underground tanks) to mid 1996 (ISO 14000). They also represent policy initiatives at different scales of aggregation: specific to a firm (underground tanks), specific to chemical industry (Responsible Care), specific to manufacturing firms across industries (33/50), and impacting virtually all firms in the economy (ISO 14000).

Following King, Keohane, and Verba (1994), I have selected the cases to ensure variations on independent variables. They also advise that with a small sample size, researchers should consciously ensure variations on the dependent variable as well. As a result, the book also examines four cases of non-adoption (ISO 14000 in Baxter and Lilly; external audits in Lilly, and Phase I of Responsible Care in Lilly) though it primarily focuses on cases where Type 2 policies were adopted.

Organization of the book

The study is organized into five chapters, including this introductory chapter. Chapter 2 presents the theoretical foundation, focusing on the new institutionalist perspective and power-based and leadership-based theories of firms. New-institutionalists focus on two broad sets of questions. First, how do institutions evolve and, second, how do institutions affect collective outcomes? This book focuses on the first question only: how do Type 2 policies – a specific genre of institutions in the context of firms – evolve and how can power-based and leadership-based policies explain their selective adoption? Chapter 3 provides a brief overview of the activities of Baxter and Lilly, and describes and compares the evolution of their environmental programs from 1975 to mid 1996. Chapter 4 examines ten cases of Type 2 policies and explores processes leading to their adoption or non-adoption. These processes are examined by employing power-based and leadership-based theories. Chapter 5 discusses the theoretical and policy implications of this book, its limitations, and issues for future research.

2 Environmental policymaking within firms

This book seeks to understand why firms selectively adopt Type 2 policies. At a broader level, it examines the evolution a specific genre of institutions – firm-level environmental policies. To explore the research question, it departs from neoclassical economic theory and employs a new-institutionalist perspective. Instead of treating firms as unitary actors, it views them as composites consisting of multiple managers. The *ultimate* unit of analysis is the individual manager within the firm.

Managers may have different preferences for Type 2 environmental policies: some – policy supporters – prioritize environmental objectives over a simple effort to maximize quantifiable profits while others – policy skeptics – do not. Once we acknowledge that some managers may pursue environmental objectives that do not or cannot meet the (quantifiable) profit criteria, firms' policies can no longer be explained as their passive responses to external stimuli, in particular, market signals and governmental regulations. Rather, policies *also* reflect intra-firm dynamics and it is difficult to predict *ex ante* whether policy supporters or policy skeptics will always prevail. The final outcome depends on a variety of factors such as policy supporters' hierarchical position, their persuasive or canvassing abilities, their expertise in the issue area, and how they invoke external factors to shape perceptions of policy skeptics. Policy outcomes are also influenced by the degree of organizational change required for the policy implementation: the greater the predicted change, the stronger the incentives are for the "losers" to oppose policy adoption. Consequently, the likelihood of policy adoption decreases.

To understand the intra-firm processes for Type 2 policies, I employ the theoretical tools of new institutionalism and integrate them with insights from sociological institutional theory and stakeholder theory. New-institutionalists emphasize the role of a broad set of institutions in shaping individual incentives, thereby impacting collective action (Bates 1983; Arrow 1987; North 1990; Ostrom 1990; Eggertsson 1990; Furubotn and Richter 1991). In the new-institutionalist tradition, I identify two relevant theories of firms: power-based, and leadership-based.[1] Though these theories view firms as composite actors, they identify different processes of internal policymaking. Individually, both theories illuminate specific dimensions of the research question but neither explains all aspects.

New institutionalism

New institutionalists integrate insights from neoclassical economics, historical institutional economics, political science, and sociology by examining the dynamics between economic and political behaviors in a given institutional context. The central questions addressed by new-institutionalists are twofold: first, how do institutions evolve in response to individuals preferences, endowments, and strategies; and second, how do they impact collective-level outcomes? *The focus of this book is on the former: why and how specific kind of firm-level institutions – Type 2 policies – evolve.*

New institutionalists distinguish institutions from organizations. Institutions are enforced rules, formal and informal, about what actions are required, prohibited, or permitted (Ostrom 1986) while organizations are collections of physical actors (North 1990). For example, courts of law are organizations, whereas laws themselves are institutions. Similarly, a firm is an organization while its policies and the rules governing the actions of its managers are institutions. Managers (unitary actors) of firms (composite actors) operate within constraints set by the institutions that are both internal and external to firms. Since institutions are the mediating variables between individuals and collectivities, new-institutionalists analyze various realms of collective action (markets, firms or commercial bureaucracies, non-profit bureaucracies, and governmental bureaucracies) using the same theoretical perspective.

The fact that institutions are influential does not necessarily imply that actors passively respond to "structures." Unlike Hayek (1955) who treats institutions (specifically, markets) as spontaneous outcomes, new institutionalists view institutions (structures) as purposive artifacts. Even though institutions matter, agents can potentially create or modify some structures at one point in time which, in turn, constrain their behavior at another point in time. For this reason, Snidal (1995) refers to institutions as endogenous parameters. As a new institutionalist, I focus on induced

[1] As explained in the introductory chapter, the third theory – transaction cost – is not considered because it is not relevant to the research question.

and imposed cooperation regarding Type 2 policies and impediments to it.

In undertaking institutional analysis, new institutionalists employ three assumptions: methodological individualism, pursuit of self-interest, and bounded rationality. These are discussed below.

Methodological individualism

Individuals are viewed as the *ultimate* units of analysis. Buchanan (1962) emphasizes the distinction between individualism as a method of analysis and individualism as a norm for organizing society. The former suggests a focus on the individual as a decision maker: what choices he/she faces and what opportunities he/she has to solve its problems.[2] Consequently, methodological individualists do not reify collective entities such as firms or governments; they treat them as composite actors. Jensen notes that:

> Organizations do not have preferences, and do not choose in the conscious and rational sense that we attribute to people. Anyone who has served on committees understands this fact. Usually no single person on a committee has the power to choose the outcome, and the choices that result from committee processes seldom resemble anything like the reasoned choice of a single individual. (1983: 327)

Methodological individualism is a key pillar in my attempt to understand policymaking in firms. I trace the evolution of Type 2 policies (institutions) of firms (organizations) to preferences, endowments, and strategies of managers. As a result, I view Type 2 policies as purposive artifacts, and institutional change (policy adoption or non-adoption) as a process that is affected by the extant structures as well as the preferences, endowments, and strategies of managers (Schelling 1978; Arrow 1987; North 1990; Ostrom 1996). Actors, however, cannot always modify or create new institutions as institutional change is affected by a variety of factors such as path dependency, transaction costs, and the rules used to change rules. Thus, and not surprisingly so, policy supporters fail to have their firms adopt Type 2 policies.

Both the neoclassicalists and the new institutionalists assume methodological individualism. However, the former contend that markets evolve

[2] Buchanan notes: '[I]ndividualism *as a method of analysis* and individualism *as a norm for organizing society* . . . [the former] suggests simply that all theorizing, all analysis, is resolved finally into considerations faced by the individual person as a decision-maker . . . analysis reduces to an examination of his choice problem and of his means or opportunities for solving this problem. To this approach is opposed that which starts from the presumption that some unit larger than the single person . . . is the entity whose choice problems are to be examined' (1962: 315).

spontaneously. The Austrian School, particularly Hayek (1945), also rejects "constructive rationalism"; he suggests that institutions cannot be constructed based on a rational design. Evolutionary economists also do not treat institutions as purposive artifacts (Nelson and Winter 1982). Similarly, sociologists or the interpretivist institutionalists contend that institutions evolve exogenously to the actors (Keohane 1988; Ostrom 1991). Thus, new institutionalists view some institutions as potentially purposive artifacts while interpretivists do not.

Pursuit of self-interest

Actors, alone or in groups, are assumed to pursue their self-interests subject to institutional constraints.[3] Both the policy supporters and the policy skeptics are self-interested actors serving their respective material and non-material (ideological or moral) objectives. Williamson (1964) argues in one of his earlier works that managers maximize utility functions that include variables such as status, salary, and prestige. It is conceivable that environmental managers have a vested interest in having Type 2 policies adopted since this often leads to increased budgets for their department and increased headcounts. This, in turn, creates promotional opportunities for them and also increases their prestige within the organization.

However, actors may not always succeed in gaining the highest level of joint net benefits and institutions may therefore generate pareto-inefficient equilibria, the Prisoners' Dilemma game being an eloquent example. Institutions may also serve the interests of dominant actors and represent distributional conflicts (Libecap 1989; North 1990; Knight 1992).

Bounded rationality

Actors are assumed to be boundedly rational: "intendedly rational but limitedly so" (Simon 1957: ix). Bounded rationality reflects information scarcity at the level of an individual actor. In the study of social systems, Eggertsson (1996) identifies three types of information issues: (1) prob-

[3] Principal–agency notions are critical in understanding behaviors of composite actors (Berle and Means 1932; Manne 1965; Ross 1974; Jensen and Meckling 1976). This book does not examine principal–agency issues relating to Type 2 policies. Also, some scholars do not view principal–agent conflicts to significantly impact working of firms. Manne (1965) argues that markets for managers substantially control agency costs. Demsetz and Lehn (1985) cite evidence suggesting that organizations with widely dispersed ownership (high principal–agent conflicts) as well as with concentrated ownership (low principal–agent conflicts) have comparable levels of profits.

lems in creating and gathering information; (2) neurological constraints in absorbing and processing information; and (3) imperfect models of the external world that limit useful applications of processed information. The first is a systemic constraint since it is practically impossible to collect perfect and complete information on every subject. In contrast to issue (1), issues (2) and (3) signify individual-level constraints; therefore, I am of the view that only (2) and (3) together reflect bounded rationality.

For Simon, bounded rationality implies that actors use rules-of-thumb or standard operating procedures (such as capital budgeting) to make decisions in repetitive situations, and *satisfice* rather than *maximize*. Although satisficing may correctly describe human behavior, it is difficult to operationalize. Some new institutionalists therefore interpret bounded rationality as utility maximization under transaction cost constraints (Eggertsson 1990).

Since boundedly rational actors may make decisions under uncertainty, decision-making may be influenced by managerial interactions. When provided with new evidence or alternate interpretations for existing evidence by policy supporters, skeptical managers may revise their assessment of policies. For example, opposing managers may change their assessment of a policy (they may begin to view it as Type 1 instead of Type 2) if a credible actor were to present the existing cost or revenue data.

Nature of goods and services

Adam Smith (1970) viewed the "invisible hand" as a mechanism to harmonize individual and collective interests when exchanges of private goods and services are organized in competitive markets. However, markets may fail when substantial externalities are present. In dealing with such market failures, individual and group rationality may conflict, leading to collective action dilemmas or social traps (Olson 1965; Hardin 1968; Platt 1973; Cross and Guyer 1980; Ostrom 1990). If actors pursue their short-term self-interests, the group arrives at a socially non-optimal situation. A socially optimal outcome can be achieved only if actors suppress their short-term self-interests.

New-institutionalists trace collective action dilemmas to the physical and institutional nature of goods and services (henceforth referred to as products). Products may initially be classified along two attributes: rivalry or subtractability and excludability (Ostrom and Ostrom 1977). Rivalry implies that if A consumes a particular unit of a product, B cannot. Some products may exhibit congestibility: non-rivalry if the number of consumers is below a given threshold, and rivalry if it is beyond that threshold. Excludability implies that it is both technologically feasible and economical for A to exclude B from appropriating benefits once a product has

Table 2.1. *The nature of goods and services*

Excludability Rivalry/subtractability	Easy	Difficult
Rival	(1) Private goods	(4) Common-pool resources
Partially rival or congestible	(2) Club goods (3) Toll goods	
Non-rival		(5) Pure public goods

Source: Adapted from Ostrom and Ostrom (1977: 12).

been produced. Excludability, reflecting the ease of defining and enforcing property rights, is influenced by both physical and institutional factors. As illustrated in table 2.1, based on physical and institutional contexts, products can initially be classified into stylized categories.

Pure public goods are non-rival and it is difficult to exclude actors from appropriating their benefits. In contrast, private goods are rival and easily excludable. Common-pool resources are rival, but it is difficult to exclude actors from appropriating their benefits. Club and Toll goods are partially rival (congestible) and easily excludable.

Public goods are specific to the jurisdictions that provision them. We can think in terms of local public goods (street lights), national public goods (national defense), and global public goods (protecting stratospheric ozone). However, public goods may create externalities for jurisdictions that are not provisioning them.

Categorizing products is important since these categories represent different types of collective action dilemmas. If actors cannot be excluded from partaking in the benefits of a product, they have few incentives to contribute to the product's provision. Difficulties in exclusion create incentives for "free riding" (Olson 1965) leading to under-provisioning of the product. Since markets elicit contributions towards provision through the threat of exclusion, they are ineffective in providing non-excludable products, and some alternative institutional mechanism is required (Ostrom 1990).

Rivalry in consumption leads to a different category of problems; it creates incentives for overuse and rent dissipation (Ostrom, Gardner, and Walker 1994). If a product is highly rivalrous and easily excludable (a private good), then its over-consumption is checked through a rise in its price. However, if exclusion from a highly rivalrous product (a common-pool resource) is difficult or costly, markets do not function effectively, and scarcity does not translate into a higher price. Unless there are counter-acting institutions, there is over-consumption leading to degradation of the resource. Again, non-market institutions are necessary to ensure sustainable use of such resources.

Scholars often assume that governmental provision is necessary for non-rival/congestible goods. However, congestible and easily excludable goods (impure public goods) can be provided by non-governmental organizations as well (Cornes and Sandler 1996). Such goods are of two kinds: toll and club.[4] The former can be provisioned by levying a user toll. Consumers, paying for every additional unit, reveal their preferences, and one can conceivably think of a continuous demand schedule for each consumer.

In contrast to toll goods, we cannot price the discrete consumption units of club goods. As a result, their provision is financed by membership fees. Industry-level initiatives such as the Chemical Manufacturers Association's (CMA's) Responsible Care (chapter 4) and ISO 14000 (chapter 4) are examples of club goods because one cannot price the discrete units of "goodwill" (Commons 1968: 199) benefits for the firms that are generated by such initiatives. Firms will have incentives to pay their membership fees only if such benefits are made excludable. Membership fees can take various forms. Consider the case of ISO 14000. Here the membership fee includes the costs of implementing environmental management systems and hiring experts to certify them. ISO 14000 certification is the "fee receipt" conferring the goodwill benefits. For Responsible Care, the membership fee takes the form of implementing new organizational policies such as community outreach programs. Since the CMA makes it mandatory for members to subscribe to Responsible Care, membership to the CMA constitutes the fee receipt.

Type 2 policies such as Responsible Care and ISO 14000 create several joint benefits. First, by reducing pollution, they create benefits for citizens at large. Since these benefits are akin to public goods, firms cannot charge for their provision. Second, these policies generate goodwill towards firms among regulators, citizen groups, local communities, and financial institutions. In addition to difficulties in monetizing them, these benefits have the characteristics of public goods. Consequently, firms have few incentives to provide them.

"Green products" (chapter 4) may also represent firms' attempts to transform benefits of environmental protection from public to private goods. If consumers are willing to pay a premium for green products (and have the satisfaction of being environmentally responsible), then firms can offset additional costs of adopting Type 2 policies. However, firms do not adopt this route *en masse* since it transfers collective action dilemmas from firms to consumers. Rational consumers may want to enjoy benefits

[4] The concept of an "impure public good" was popularized by Buchanan (1965) though its application can be found in works of Tiebout (1956), Wiseman (1957), Pigou (1960) and Olson (1965). I am indebted to Brenda Bushouse for highlighting the distinction between club and toll goods.

of a clean environment (from which they cannot be excluded) without paying for them. If such defections are widespread, then markets for premium green products would remain small, and firms pursuing profits would have few incentives to adopt Type 2 policies.

Type 2 policies also generate goodwill for firms. To transform non-excludable goodwill benefits to club goods, industry-level institutions may be required. Initiatives such as the Environmental Protection Agency's (EPA) 33/50 program, and the CMA's Responsible Care (all chapter 4) fit into this category. Through Responsible Care, the CMA seeks to generate goodwill for the chemical industry, particularly for its members. To ensure that its members do not free-ride, the CMA *requires* that all its members subscribe to Responsible Care. Similarly, firms adopting ISO 14000 management systems will get the ISO certification conferring goodwill benefits to them. This again transforms goodwill benefits from a public good to a club good.

Goodwill benefits could also be transformed to private goods. For example, financial organizations (such as mutual funds, insurance companies, and pension funds) may reward "green firms" that adopt Type 2 policies. However, as Schmidheiny and Zorraquin (1996) point out, since the primary objective of financial institutions is to maximize shareholders' wealth, most of them would probably penalize firms that adopt Type 2 policies. Collective action dilemmas are now at the investor level: rational investors may desire the benefits of a clean environment (from which they cannot be excluded) without paying for them. Since "green investors" constitute only a niche market for investment funds investing in "green securities" only, "green funds" remain peripheral players in financial markets.

Notions of efficiency and capital-budgeting

The concept of efficiency is intrinsic to neoclassical economic theory. Markets are viewed to lead to efficient macro-level outcomes while profit-maximizing firms are viewed to work towards maximizing the efficiency of their operations. The term efficiency has many interpretations and I discuss two relevant notions below. My argument is that no matter which definition of efficiency we adopt, neoclassical theory cannot explain why firms selectively adopt Type 2 policies (Prakash 1999b).

Substantive and procedural efficiency

Since neoclassical theory views firms as profit maximizers, the issue arises as to how do firms maximize profits in their day-to-day activities. Drawing on Simon's (1957) notion of substantive and procedural rationality, this book distinguishes between two notions of efficiency: substantive and procedural.[5] The former suggests *what* policies ought to be adopted while the latter describes *how* they should be adopted.

Theories employing the notion of substantive efficiency assume that managers are fully rational, having full information about the future costs and benefits of a policy. They, therefore, focus on policy outcomes rather than on the processes leading up to them. Substantively efficient firms adopt only those policies that *ex ante* maximize quantifiable profits. What if, as in the case of Type 2 policies, managers do not or cannot quantify profits? Or, if some policies are viewed to serve non-profit objectives? Clearly, any theory based on substantive efficiency offers little help in this regard. It could perhaps be argued that substantive efficiency is an instrumental criterion; this is how firms ought to behave. If treated as a descriptive criterion, we should focus on systemic, not firm-level, outcomes. In the long run, at a systemic level, only those firms that employ the notion of substantive efficiency will survive (Alchian 1950).[6] This may well be correct and I therefore do not dispute the usefulness of substantive efficiency for certain research questions. However, this notion is not helpful in answering my research question that focuses on firms (not the system) and examines the processes leading to adoption or non-adoption of a given Type 2 policy.

I now turn to the notion of procedural efficiency. Instead of focusing exclusively on the final outcome, procedurally efficient managers employ procedures that seek to maximize quantifiable profits. This assumes that such procedures, on average, have tended to produce substantively efficient outcomes. Simply having a procedure in place to maximize profits constitutes "intended efficiency" only; evidence of their success, on average, makes them procedurally efficient. On this count, substantive and procedural efficiency are two facets of the same managerial objective: maximization of quantifiable profits.

Capital budgeting

Since this study focuses on intra-firm dynamics and procedures regarding Type 2 policymaking, the notion of procedural efficiency is the more appropriate to examine whether policy outcomes satisfy the profit maximization criteria. I operationalize procedural efficiency in terms of standard and well-accepted methods of project appraisal, often subsumed under

[5] Simon suggested the notion of procedural rationality as a critique of neoclassical theory's notion of substantive rationality.

[6] Alchian's contention is perhaps not falsifiable in that it does not stipulate any *ex ante* criteria.

capital budgeting.[7] To cope with uncertainty about future costs and revenues, and to impart objectivity to investment analysis, managers adopt procedures such as capital budgeting. Predicting profits is difficult due to the bounded rationality of managers as well as uncertainty about variables exogenous to firms. Baumol describes capital budgeting as follows:

> Capital budgeting refers to investment decision-making procedures of business firms and other enterprises. The subject encompasses such topics as the selection of projects (which new factories, if any, should the company build), the timing of investment, the determination of the amount to be invested within any given time period, and the arrangements of the financial means necessary for the completion of the project. The calculations which are appropriate for these decisions for the most part derive directly from the theory of capital. . . . However, there is an extremely important limitation which must be emphasized from the very beginning. Imperfect foresight into the future, risk, and uncertainty will for the most part be ignored because economists have not devised really effective methods of taking them into account in the analytical methods. Unfortunately, capital-budgeting is the one subject were we can least afford to abstract from limitations in our knowledge of the future, because, by its very nature, the investment decisions can only be justified in terms of their prospective effects. (1982: 597)

As a tool for investment analysis, capital budgeting (or discounted cash flow techniques) is well established in the Anglo-American system of capitalism (Schall, Sundem, and Geijsbeek 1978; Wicks 1980; Oblak and Helm 1980). The two popular capital budgeting methods are the Net Present Value (NPV) method and the Internal Rate of Return (IRR) method. In the NPV method, managers make projections of costs and revenues emanating from the project over a ten to twenty year period, and discount these cash flows by firms' cost-of-capital. Broadly, cash flows equal net profits plus depreciation. Depreciation is added back to the equation since it is only an accounting cost, and not a cash outflow. Cost-of-capital is the cost of raising funds in the capital market; alternatively, the opportunity cost of investing funds in the next best alternative. If the net present value of the discounted cash flows is positive, such projects are deemed as being potentially profitable, and, as a result, firms consider investing in them.

In employing the IRR method, managers identify the discount rate that equates discounted benefits to discounted costs. If this discount rate is greater than the firms' cost-of-capital (also called the hurdle rate), firms consider projects investment worthy. The NPV and the IRR methods are similar but not identical. Specifically, in the IRR method there can be multiple discount rates equating discounted benefits to costs.

[7] In public finance literature, capital budgeting refers to an accounting procedure and not an investment appraisal procedure. In this book, capital budgeting refers to the latter only.

Capital budgeting is important because managers put forward their best estimates of benefits and costs emanating from a particular project. Such procedures enable them to cope with uncertainty, and to establish transparency and impartiality in the project appraisal process. Financial markets perceive this as evidence of firms' commitment to financial discipline. With active transnational mergers and acquisitions markets, managers have incentives to closely monitor the reaction of financial markets to firms' investment decisions. Further, in many firms, compensation of top executives is linked to the price of a company's stock (the deferred payment compensation). Hence, managers have incentives (that is, the "stick" of being taken over by another firm, and the "carrot" of enhancing their earnings) to closely attend to the evaluation of financial markets about their firms' performances. Finance and accounting managers, in particular, are major supporters of employing capital-budgeting to assess project viability. Deviations from this practice are rare, especially if projects involve millions of dollars.

Arguably, there are other methods of investment analysis as well such as full-cost accounting and life-cycle analysis that firms could potentially employ. These methods force firms to take into account long-term social (as well as private) costs, thereby internalizing environmental externalities. I have not examined these alternative investment appraisal techniques for several reasons. First, Lilly and Baxter do not routinely employ any of these methods while they do employ capital budgeting. Further, even as a non-routine measure, Lilly and Baxter did not employ them in making decisions on Type 2 policies that this book examines. Second, the methodologies and operationalization of these techniques are not standardized. Hence, they lack legitimacy, especially with the finance and accounting managers. Third, these techniques may not be compatible with the dominant paradigm about the nature of the firm – the main objective of firms is maximizing shareholders' wealth, an objective that capital budgeting focuses on. This study is arguing that this dominant paradigm is inadequate to explain selective adoption of Type 2 policies. Hence, the existence of alternative methods that are not standardized, that lack wide-spread acceptability within and across firms, and that neither Lilly nor Baxter employ, does not undermine my operationalization of procedural efficiency in terms of capital budgeting.

Capital budgeting is indeed a well-established and widely employed tool of investment analysis which reinforces the argument that firms and their managers attempt to be procedurally efficient. As a result, any deviations from these established procedures (such as adopting Type 2 policies) is puzzling and worthy of examination. The book undertakes this task by "unpacking the firm" and linking policy outcomes to organizational dynamics and politics.

Firms as composite actors

Neoclassical economics does not explain an important puzzle: why do firms arise at all? That is, why and how do managers arrive at "make or buy" decisions? Coase's (1937) seminal article, "The Nature of the Firm," emphasized transaction costs as the key factors in the emergence of firms. Coase criticizes the neoclassical theory in the following terms:

> [It assumes] that the consumer is not a human being but a consistent set of preferences. The firm is defined as a cost and a demand curve. The theory is simply a logic of optimal pricing and input combination. Exchanges take place without any specification of its institutional settings. (1988: 2)

In his Nobel Prize lecture, Coase further observed that:

> [A]n economist does not interest himself in the internal arrangements within organizations but only in what happens on the market, the purchase of factors of production and the sale of the goods and services that these factors produce. What happens in between the purchase of the factors of production and the sale of goods that these factors produce is largely ignored. . . . *The firm in mainstream economics has often been described as a "black box." And so it is.* This is most extraordinary since most of the resources in a modern economic system are employed within firms, with how these resources are used depends on administrative decisions and not directly on the operation of the market. Consequently, the efficiency of the economic system depends to a considerable extent on how these organizations conduct their affairs, particularly, of course the modern corporation. Even more surprising, given their [neoclassical economists'] interest in the pricing system, is the neglect of the market or more specifically, the institutional arrangements to a large extent what is produced, what we have is a very incomplete theory. (1993: 228–9; italics mine)

Coase took the first step in "unpacking the firm." Williamson (1985), building on Coase's ideas, provides a more developed theory to understand why transaction costs arise, and how they impact managerial "make or buy" decisions. Williamson emphasizes incomplete contracting since it is impossible to account for every contingency. He assumes that humans are potentially opportunistic; they pursue "self-interest with guile." He highlights the "fundamental transformation" (from competitive markets to a bilateral monopoly) in many repeated interactions due to "asset specificity" – the productivity of an asset is specific to certain persons and tasks. Asset specificity makes asset owners vulnerable to labors' opportunism, and purchasers and suppliers vulnerable to each other.

Controlling opportunism through external enforcement ("legal centralism") is costly; actors, therefore, create hierarchies to control it internally at lower transaction costs. Thus, firms are hierarchical governance structures that manage interdependence among economic actors.[8]

Coase, Williamson, and other scholars in the transactions cost tradition have highlighted the shortcomings of neoclassical economics and contributed key concepts for unpacking the firm.[9] Transaction cost theories focus on macro-institutional issues such as why firms arise. They are, therefore, inappropriate for this study since it focuses on institutional evolution at a micro-level – why firms adopt Type 2 policies selectively. To answer this question, the book employs power-based and leadership-based theories, both within the new-institutionalist tradition, to "unpack the firm" and to examine intra-firm dynamics on Type 2 policies.

Power-based theories

The term "power" has many meanings. This book employs it to describe the ability of manager A to influence outcomes in wake of opposition from manager B. For example, policy supporters are deemed to exercise power if they force adoption of Type 2 policies although policy skeptics oppose it (Weber 1947; Dahl 1957; Pfeffer 1981; Mitchell, Agle, and Wood 1997).

It is instructive to examine the bases of power and why some managers are more powerful than others. Etzioni (1988) identifies three types of power – coercive, material, and symbolic. Coercive power is based in control over instruments of coercion. For example, some managers may control workers by physically intimidating them. Materially powerful managers control instruments of material well being. For example, a supervisor can control compensation and promotions of a subordinate. Symbolic power suggests that managers control normative symbols that

[8] Chandler (1962, 1977) also views modern industrial enterprises as efficiency-enhancing governance structures. His notion of transaction costs is restrictive (or more precise) compared to that of Williamson's. Chandler focuses only on the reduction in administrative costs of coordinating resource flows in manufacturing when the production technology permitted mass production of standardized products for national as well as international markets. Further, Chandler does not relate transaction costs to opportunism due to asset specificity.

[9] Williamson is criticized for inadequately explaining how managers operationalize asset specificity and opportunism (Dugger 1983; Dow 1987; Roberts and Greenwood 1997). Critics contend that he provides insufficient guidance for measuring transaction costs. As a result, it is difficult to predict *ex ante* whether managers would favor "make" or "buy" for a specific policy. Critics also suggest that as a key behavioral assumption, opportunism is a bad descriptor and predictor of human behavior. It may also turn out to be a self-fulfilling prophecy (Ghoshal and Moran 1996).

bestow prestige. For example, a supervisor may have the ability to decide on the designations of his/her subordinates. Or, the supervisor could control allocation of work responsibilities; the favored ones could work on prestigious and high-visibility projects. For the purpose of this study, material and symbolic bases of power are relevant. Hierarchically superior managers typically have symbolic and material power over subordinates. Type 2 policies are adopted by a power-based route if policy supporters are either hierarchically superior or can capture the top management, thereby imposing it on policy skeptics. Similarly, hierarchically senior policy skeptics can ensure that Type 2 policies are not adopted.

Power-based approaches are well established in the new-institutionalist perspective. New institutionalists point out that institutions may not always arise to internalize efficiency gains; they may represent distributional conflicts (Libecap 1989; North 1990; Bowles and Gintis 1993). Knight observes that:

[T]he ongoing development of social institutions is not best explained as a Pareto-superior response to collective goals or benefits but, rather, as a by-product of conflicts over distributional gains. (1992: 19)

Similarly, scholars question if firms are merely institutions for organizing value-addition processes. Or, as radical scholars suggest, they are instruments to dominate labor (Perrow 1979) and to facilitate accumulation (Marglin 1974). Edwards (1979) acknowledges the coordination function of firms. However, as coordination can be achieved by tradition and peer groups, he views firms more as instruments for bureaucratic control. Putterman (1984) questions the efficiency rationale for the emergence of firms because labor hiring capital has the same efficiency implications as capital hiring labor.

If firms are organizations to serve the interests of dominant actors, policies of firms should reflect domination of these interests. Adopting Type 2 policies through a power-based route is predicted to generate conflict. Policy skeptics will accept such policies not because they buy into their logic; rather, their acceptance represents the victory of hierarchical superiors (policy supporters) over subordinates. Policy skeptics are not predicted to change their preferences (as reflected in their behavior) about the desirability of such policies. The level of opposition is influenced by factors such as their extent of disagreement with the policy and the fear of retribution. As I elaborate in chapters 4 and 5, if a policy is perceived to reduce departmental budgets, headcounts, or managers' organizational clout, the "losers" could be expected to assume the role of policy skeptics. In such instances, power-based processes are perhaps the only route to have such policies adopted. Power-based processes are therefore predicted to generate covert and overt intra-firm conflicts. Opposing managers may capitulate, but still carry on guerilla warfare. If a major project is adopted that does not satisfy capital budgeting and hurts the interests of mangers, I examine evidence of contestation. Policy skeptics may exhibit their disagreement in a variety of ways such as slowing down policy implementation or even openly sniping at policy supporters during meetings; they may greet reports on policy failures with cynical "I told you so" remarks. I observed such behaviors during my one-to-one interviews as well as in business meetings which I attended as an observer.

Leadership-based theories

Leadership is a highly researched issue in organizational theory.[10] Similar to power-based theories, leadership-based theories also suggest that certain managers play key roles in creating or modifying institutions (Barnard 1938; Follett 1940; Boulding 1963; Miller 1992). Unlike the dominant actors in power-based processes who impose their preferences, leaders are consensus inducers. They have both the political savvy and yet, more ennobling and ethical goals (Lipman-Blumen 1996). Importantly, policy consensus may not arise spontaneously (as in the Hayekian notion of spontaneous cooperation); interventions of leaders are required. Thus, leaders have the ability to build a shared vision and to foster systemic and long-term patterns of thinking through dialogue (Selznick 1957; Senge 1994; Weick 1995).

Leadership-based theories suggest that the presence of leaders is essential for firms to arise and function.[11] This perspective of the nature of firms contrasts with Williamson's who views hierarchies as artifacts to economize on costs of labors' opportunism. Williamson's critics argue that managers may also behave opportunistically (hierarchical failure) by unfairly appropriating profits (Kreps 1990). Firms cannot, therefore, be viewed simply as artifacts to mitigate market failures; the role of leadership is important. Miller notes that:

[10] Leadership theories can be classified into the following categories: Trait theories (Reitz 1987; Katz 1974; Yukul 1981); Group and Exchange theories (Greene 1975; Graen, Novak, and Sommerkamp 1982); Contingency theories (Fiedler 1967); Path-Goal theories (Georgopolous, Mahoney, and Jones 1957; House 1971); Charismatic leadership theories (House 1976; Bass 1985); Transformational leadership theories; and Social Learning theories (Davis and Luthans 1980). For a survey of literature on leadership see Chemers and Ayman (1993), Luthans (1995), and Northouse (1997).

[11] The influence of leaders as agents of change is not limited to the private sector. There is a growing literature on policy entrepreneurship in the public sector (Schneider and Teske with Mintrom 1995).

Managers face short term-incentives to choose inefficient incentive regimes for subordinates. Employees, knowing this, have no reason to trust the employers with information that would make it possible for the employers to make inefficient decisions. Hierarchy is thus a setting for a commitment problem. *The problem can be solved, but only through a set of strategies that are essentially political . . . [and a] leadership style projecting trustworthiness* and/or constitutional constraints on the political authority of the hierarchical superiors. (1992: 235; italics mine)

Barnard (1938) suggests that managerial functions extend beyond organizing factors of production, devising appropriate incentive structures to minimize shirking, and writing contracts to guard against opportunism of the subordinates. The distinctive managerial function is to motivate the employees to surmount their narrow, short-term self-interest. Lipman-Blumen (1996) identifies three general styles employed by leaders: direct (managers tightly define their goals and achieve leadership by outstanding performance), relational (managers lead by collaborating, contributing, and empowering people to achieve respective individual goals), and instrumental (managers employ personal relationships and organization politics to achieve their goals, while allowing others to shape the pathways to those ends). This book focuses on instrumental leadership style. Intra-firm dynamics in adopting Type 2 policies are leadership based if they reflect a conscious building of consensus by policy supporters. Since these policies do not increase quantifiable profits, or in some instances create "losers" from organizational change, they could initially be opposed by policy skeptics (or indifferently received by policy neutrals). Due to inter-managerial conflict, policies are often either shelved or adopted by top management imposition (a power-based processes). However, there is a third route as well. Policy supporters may repackage such policies to highlight the long-term benefits, remove the objectionable but unimportant aspects, and persuade policy skeptics to revisit their assessments of benefits and costs. Instead of relying on quantifiable profits as the sole criterion, policy supporters may suggest employing multiple criteria to assess the desirability of policies.[12]

[12] In terms of Follett (1940), this could be termed as "integration." According to her, issues of conflict can be dealt with by domination (similar to power-based processes), compromise (achieving middle ground without changing preferences), and integration (involves changing for preferences). Conflict could be constructive, if employed to achieve integration. In my discussion of leader-based processes, integration is being achieved since policy skeptics revise their preferences on the desirability of Type 2 policies. Unlike Follett's conception where both parties change their preferences (what she calls "reciprocal adjustment"), I focus on changing preferences of policy skeptics only.

To employ Boulding (1963) terminology, a leadership-based process could be classified as an integrative response: "[W]hich establishes a community between the threatener and the threatened and produces common values and common interests" (p. 428). In the context of this book, it could be interpreted as a policy outcome that establishes a shared understanding between policy supporters and policy skeptics on the desirability of a Type 2 policy.

How are agendas transformed, deadlocks broken, and what is the role of leaders in this context? Ikenberry's (1993) study of the evolution of the Bretton Woods economic system is instructive on this count. According to Ikenberry, the interventions of policy entrepreneurs at crucial junctures in the form of new ideas and new ways of conceptualizing interests breaks deadlocks, thereby providing a basis for building of consensus. He notes that:

Miracles aside, how does one explain the Anglo-American postwar settlement? Can a simple interest-based argument explain the settlement, or do we need to probe more deeply into the manner in which interests were defined, coalitions were forged, and power was legitimated in the aftermath of world war? I argue that we must probe deeply. . . . A set of policy ideas inspired by Keynesianism and embraced by groups of well-placed government specialists and economists was crucial in defining government conception of postwar interests, building a coalition in support of the postwar settlement . . . these experts and their "new thinking" were important in overcoming political stalemate both within and between the two governments. (1993: 58)

It appears that leaders adopt a variety of strategies to influence policy skeptics. As I discuss in chapters 4 and 5, these strategies often involve impressing the long-term benefits of Type 2 policies upon the skeptics. The presence of a coercive external stakeholder (such as the EPA) that can impose excludable costs or bestow excludable benefits encouraging such policies, imparts credibility to their arguments. Leaders also rely significantly on communicating and sharing information, a tactic that has effectively been used by policy supporters in Baxter. Providing new information and framing existing information in new contexts may stimulate policy skeptics to reassess their preferences and the benefits of a Type 2 policy. They may also seek to create an organizational culture of environmentalism through devices such as regular newsletters and conferences.

The personal credibility of leaders as consistent champions of particular issues, along with their expert knowledge about the issue area, is important.[13] If policy supporters do not have sufficient professional expertise, they could utilize outside experts to make presentations to policy skeptics on the desirability of adopting a policy. Thus, leaders could adopt a variety of strategies to influence the skeptics to revise their assessments of Type 2 policies. I do not expect policy skeptics to change their positions overnight; sustained canvassing by policy supporters is often required. During my one-to-one meetings, one manager admitted that some of them were initially very skeptical about "throwing money

[13] Also, the role of technical expertise of bureaucrats in influencing public policy outcomes is well established (Lowi 1969; Allison 1971; Dodd and Schott 1979; Katzmann 1980; Khademian 1992)

down the drain" but the "tree huggers do have a point; this benefits us in the long-term."

In both power-based and leadership-based processes, policy supporters are expected to portray Type 2 policies as serving long-run profit and non-profit objectives of their firms. Although policy supporters may make claims about increases in long-term profits, they provide no quantifiable estimates. Profits no longer remain an "objective" concept whose measurement is invariant across actors. This study is not arguing that established procedures of project appraisal are irrelevant. They matter very much and that is why it is difficult for policy supporters to justify why their pet policy should not be subjected to the formal rules of project appraisal. Importantly, such exceptions occur often in evaluating environmental projects, and this study proposes one way of examining the processes that lead to such exceptions.[14]

The next chapter provides an overview of the activities of Baxter and Lilly, and describes and compares the evolution of their environmental programs from 1975 to mid 1996. It lays out the broad institutional contexts in which these firms made policies on environmental issues. A knowledge of these firms' internal and external contexts is critical for understanding the dynamics of power-based and leadership-based processes, both when they led to the adoption as well as the non-adoption of Type 2 policies. Employing power-based and leadership-based theories, chapter 4 then examines ten cases of Type 2 policymaking: four common to Baxter and Lilly (underground tanks, 33/50, ISO 14000, and environmental audits), and one each idiosyncratic to them (Responsible Care to Lilly and green products to Baxter). Finally, in chapter 5, the findings and conclusions of this book are presented.

[14] I disagree with the contention that leadership-based processes are akin to power-based processes where power is exercised subtly in terms of shaping opinion. This issue is discussed in chapter 5.

3 Baxter and Lilly: evolution of environmental programs

The previous chapter outlined a theoretical approach to "unpack" firms for studying their internal policy dynamics on environmental issues. This chapter examines the evolution of the environmental function in Baxter and Lilly. It describes and compares their internal environmental policy-making structures. Both Baxter and Lilly have come a long way in developing and institutionalizing their environmental programs. Baxter's and Lilly's experiences are fairly representative of large US firms which confronted a new set of managerial challenges in the 1970s in the health, safety, and environmental spheres. Though many challenges have yet to be tackled, it is fair to say that most firms have undergone a fundamental transformation on how they relate their functioning and policies to the natural environment. This transformation, however, is uneven within and across firms and can be attributed to factors internal and external to firms. The next chapter focuses on the key events in the evolution of Baxter's and Lilly's environmental programs. These cases are examined by employing power-based and leadership-based theories

Baxter: an overview

Baxter International Inc. is a Deerfield, Illinois-based multinational corporation that develops, manufactures, and markets products and services used in hospitals and other health-care settings. On October 1, 1996, a part of Baxter International Inc. was spun off into an independent firm called Allegiance Corporation. As shown in table 3.1 below, the 1995 net sales of undivided Baxter International Inc. (henceforth Baxter) stood at $9.6 billion, and research and development expenditure at $351 million. Since I examine environmental policymaking in the undivided Baxter International Inc. during 1975–March 1996, the splitting of Baxter into two separate firms does not impact my research design

Baxter had five major business portfolios. The first four now constitute the (new) Baxter International Inc., and the fifth has been spun off as Allegiance Corporation (Baxter 1996a):

Table 3.1. *Baxter International Inc.: Financial highlights (in millions of dollars)*

	1995	1994	Change
Net sales	9,619	9,324	3.1%
Income from continuing operations	435	596	(27.0)%
Return on shareholder's equity	n.a.	17.3%	
Research and development expenditure	351	343	2.3%
Research and development expenditure as a percentage of net sales	3.6%	3.7%	

Sources: (1) Baxter (1994a: 1).
(2) Allegiance (1997).
(3) Baxter (1997).

(1) Renal Division provides products and services for patients suffering from chronic kidney failure. Baxter pioneered hemodialysis in 1956 and is now investigating animal-to-human kidney transplants.
(2) Biotechnology Group produces plasma derivatives, biotechnology, and blood-handling products and conducts research to treat blood diseases, cancer, and diabetes.
(3) Cardiovascular Group provides products to fight late-stage heart and vascular diseases. It manufactures heart valves, valve repair products, cardiac monitoring systems, catheters, and equipment used in vascular surgery.
(4) I.V. Systems/International Hospital Divisions manufacturer and distribute intravenous solutions and related products.
(5) Hospital Management Business manufactures and distributes surgical and respiratory therapy products and offers services such as inventory management, customized packaging for surgical procedures, and identification of savings opportunities. This division has been spun off as Allegiance Corporation.

Evolution of environmental programs

Baxter's environmental policies have passed through three phases. Baxter had a state-of-the-art environmental program in the late 1970s and the early 1980s. Its environmental programs began to backslide in the mid 1980s. The late 1980s marked the beginning of the third phase when Baxter revitalized its environmental programs. The top-management mandated that Baxter's US-based facilities should establish a state-of-the-art environmental program by 1992 and facilities abroad by 1996. I elaborate on these phases below.

Baxter's environmental programs had an early start. By the mid 1970s, the senior management had recognized the importance of environmental issues, and there were discussions within the company about creating internal environmental institutions and organizations. In 1976, Baxter (then known as Baxter Travenol Corporation) established its environmental department. Since at that time environmental policies were viewed as having predominantly legal and technical dimensions, Ray Murphy and Don Nurnberg were transferred from the Legal and Corporate Facilities Engineering departments respectively to Corporate Environmental Affairs.

For three reasons, 1977 constitutes a watershed in the evolution of Baxter's environmental programs. First, Murphy and Nurnberg organized Baxter's first-ever company-wide environmental conference. This signaled to the whole organization the intent of the top management to emphasize environmental issues and to encourage the active participation of managers from its various divisions and facilities. An outcome of this conference was that environmental coordinators were appointed for all US facilities. This laid the foundation of the facility-level infrastructure for developing and implementing environmental programs.

Second, to institutionalize the momentum generated by this conference at the top management level, a committee of senior executives, the Environmental Review Board (ERB), was established to oversee and guide company-wide environmental programs. G. Marshall Abbey, Senior Vice-President and the General Counsel, was appointed the chairperson of the Board, a position he held until his retirement in 1992. Abbey, a lawyer by training, recognized the growing importance of environmental issues, and that Baxter needed to proactively establish a strong environmental program. Abbey found allies in other members of the ERB: D.G. Madsen and C.F. Kohlmeyer.

Third, the corporate environmental group developed Baxter's first environmental manual and began auditing facilities' compliance with the laws that were identified in the manual. Manuals are useful tools for auditing performance since they explicate the expectations and responsibilities of various positions in the organizational structure. Consequently, developing a manual was an important step in translating the intent behind creating new decision-making organizations (such as the ERB) into concrete policies.

Over the next few years as the responsibilities of the environmental function expanded, communication bottlenecks began to appear. Some managers felt the need to establish a source of information on new developments in corporate environmental policies; annual updates during the environmental conferences were not sufficient. Further, since new laws were being enacted at the Federal and state levels, there was also a need to have updates on how such laws may impact Baxter. As a result, in 1980, the corporate environmental group began publishing Baxter's internal environmental newsletter, *Travenol Environmental Newsletter*, edited by Ray Murphy.

The late 1970s was a good time for Baxter. Its business was booming, profits were robust, and environmental regulations were relatively uncomplicated and easy to meet. In the early 1980s, the health care industry, facing a cyclical downturn, came under severe financial pressure. In a highly controversial move, both within and outside the company, the senior management responded to this exogenous development by merging with American Hospital Supply Corporation (AHSC) in 1985. It was one of the biggest mergers of that time with both of the firms having sales of about $3 billion each. The preoccupation with this stormy merger meant that Baxter's senior managers had less time to devote to 'soft-areas' such as Baxter's environmental programs. For example, in rushing through the merger, they did not correctly appraise AHSC's environmental liabilities. AHSC did not have a strong environmental program; in contrast to eleven managers in Baxter's corporate environmental department, AHSC had only one. However, Baxter's senior management felt that the strategic importance of merging with AHSC far outweighed AHSC's environmental liabilities.

Soon after the merger, as a part of corporate restructuring, the environmental budget of the merged firm actually decreased in absolute terms. Consequently, Baxter's environmental program began suffering from inadequate human resources. This was further accentuated by attrition.

Unfortunately for Baxter, such budget squeezing and attrition coincided with the increasing number, stringency, and complexity of environmental regulations at the federal, state, and local levels. Environmental issues had begun to have a dramatic impact on the costs of doing business, especially on issues such as real-estate transactions, landfills, and transportation of hazardous chemicals. Local communities and environmental groups had also become very aggressive in monitoring the environmental performance of firms. To add to Baxter's woes, unflattering media reports on toxic emissions from its facilities began appearing, even though such emissions were legally permissible. In his address to the company's annual environmental conference, Vernon Loucks, Baxter's Chief Executive Officer (CEO) admitted that:

> But for the most part, since the mid-1980s, Baxter's environmental program has slipped. There are several reasons for that. First, there has been an explosion of environmental regulations. . . . Second, our company has gone through some major changes. Early in the 1980s, the pressure to control health-care costs began to build. . . . Our merger with the American Hospital Supply Corporation in 1985 and subsequent restructuring were necessary responses to these developments. . . . But in addressing these vital issues, the company lost sight of the importance of a good environmental program. We began to take our eye off the proactive measures needed to sustain compliance. Scores of new people were taking on environmental duties at our facilities, but we were not giving them the training and time they needed to do their jobs. (Baxter 1990b:1–2)

Prompted by negative media coverage and litigations by the Environmental Protection Agency, Baxter refocused on environmental programs in the late 1980s. Four key individuals – Senior Vice-President, G. Marshall Abbey; Vice-President, Charles F. Kohlmeyer; Head of Environmental Law, Ray Murphy; and Senior Counsel, and subsequently, Ray Murphy's successor, William R. Blackburn – saw within such external pressures the opportunity to push through a pro-environment agenda within Baxter. They found a sympathetic ear in Vernon Loucks. These managers argued that Baxter must proactively deal with environmental issues by having a state-of-the-art environmental program. Abbey and Kohlmeyer marketed this vision to the top management, and Murphy and Blackburn, to the rest of the organization.

In the late 1980s, taking initial steps towards realizing this vision, Baxter adopted three sets of beyond-compliance policies. First, in 1988, the ERB decided to remove all underground storage tanks from its facilities around the world and replace them with expensive tanks having beyond-compliance features (chapter 4). Second, in 1989, the ERB decided that Baxter should aggressively reduce its emissions of air toxins and chlorofluorocarbons (CFCs). With 1988 as the baseline, the ERB set a target for reducing these emissions by 60 percent by 1992, and by 80 percent by 1996 (chapter 4). Third, the ERB decided to assess the environmental liabilities of AHSC facilities; Arthur D. Little (ADL), a leading environmental consulting firm, was hired to develop a screening protocol (chapter 4). Based on this protocol, ADL audited twenty six facilities and reported that environmental programs were indeed in a state of neglect (Baxter 1990a).

In December 1989, Ray Murphy retired as the head of Corporate Environmental Law. As discussed earlier, Murphy contributed significantly towards institutionalizing the environmental function within Baxter. William Blackburn succeeded Ray Murphy as the head of Environmental Law Affairs, and in 1992 he was named as the Vice-

President of the newly created division of Corporate Environmental Affairs.

Soon after assuming his new responsibilities, Blackburn, along with Ron Meissen of Corporate Environmental Engineering (and their respective teams), began preparing a blueprint to revitalize Baxter's environmental programs.[1] On February 16, 1990, Blackburn presented this plan to the ERB where it was unanimously approved. Under this plan, Baxter would establish a state-of-the-art environmental program; ADL was invited to help in this task. Note that ADL had been hired in 1988 to develop a screening protocol. As I discuss in chapter 4 (environmental audits), the adoption of state-of-the-art standards on an accelerated schedule, and the invitation to ADL, was resisted by some facility- and division-level managers. However, a forceful advocacy of the new policy by senior management, especially Loucks and Abbey, left little room for such opposition to continue. As Loucks noted:

Unfortunately, business hasn't done a good job of acting on environmental issues in the past. At least that's what the public thinks. A recent survey by the Roper Organization found that only one-third of those polled felt that business was meeting its obligations to protect the environment. Companies that try to meet this obligation with words alone will not be taken seriously.

It's tempting to ask: Can't we just lie low, wait for the storm to pass, then go about business a usual? No, this won't blow over, and this is no time for our industry to display a bunker mentality. While survey results may shift from time to time, the public's high interest in the environment will remain . . . (Baxter 1992a: 1–2)

Baxter adopted a two-stage strategy. First, with help from ADL, Blackburn's team defined state-of-the-art standards for firms in similar environmental risk categories as Baxter. In 1989, Blackburn had already identified management systems adopted by facilities with strong environmental programs. The first cut of the state-of-the-art standards was based on these data. Once the state-of-the art standards were defined, ADL audited Baxter's corporate and divisional programs against these standards. Since such standards are dynamic, ADL is invited every three years to redefine these standards and recertify Baxter's environmental programs.

ADL reported that by the end of 1990, Baxter's program at the corporate level had progressed two-thirds of the way towards meeting state-of-the-art requirements; programs at the divisional levels met about 40 percent of the state-of-the-art requirements with some divisions as low as

[1] I understand from at least two sources that Bill Blackburn was spotted working on this document even on Christmas Eve!

7 percent and others at 80 percent level. Assessments done by the Corporate Environmental Group suggested that environmental programs at the facility level were about 35 to 45 percent of the state-of-the-art requirements (Baxter 1991a).

Following the ADL report and the new policy, Vernon Loucks mandated that all facilities based in the US, Canada, and Puerto Rico should achieve the state-of-the-art standards by 1993, and facilities abroad by 1996. Consequently, the ERB asked Blackburn and his team to prepare a plan for revamping environmental programs. Further, to monitor progress on various environmental initiatives, they also asked Blackburn to prepare a yearly State-of-the-Program Report for presentation to the ERB.

Following ERB's directive, many organizational and institutional changes were implemented. First, a new environmental manual was issued. This manual outlined the responsibilities, procedures, forms, and audit checklists to help environmental managers, whether at facilities, divisions, or at the corporate office, in performing their functions. Second, to give higher visibility and coherence to environmental function, the Environmental Law section was spun off from the Law Department and established as a separate function: Corporate Environmental Affairs. Further, to facilitate closer interaction between Corporate Environmental Engineering (CEE) and Corporate Environmental Affairs (CEA), their offices were relocated in physical proximity. Third, the major responsibilities to design and implement environmental programs was moved from Corporate Groups to divisional environmental managers (DEMs). As a result, new DEMs were recruited; their number increased from six in the beginning of 1990 to fourteen by the end of 1990 (Baxter 1991a).

Packaging reduction initiatives were the next landmark in the evolution of Baxter's environmental program. This was prompted by a challenge put forth by the Coalition of Northeastern Governors (CONEG) to the top 200 users and producers of packaging in the US. In May 1991, on behalf of CONEG, New Jersey Governor Jim Florio, in a letter to Vernon Loucks, asked Baxter to voluntarily set goals to reduce its packaging wastes and periodically report its progress to CONEG's Source Reduction Council. Accepting the CONEG challenge, Baxter committed to a 15 percent reduction in per-unit packaging weight by 1996 with 1990 as the baseline (Baxter 1991b). The ERB asked Corporate Engineering to work with divisions for establishing numerical goals for facilities and divisions. On this count, Baxter went beyond the CONEG challenge in that CONEG's packaging reduction guidelines did not contain any deadlines or reduction targets (Baxter 1992b). Baxter's Packaging Reduction Task Force set both target dates and numerical objectives on the following:

(1) eliminate foams made of CFCs;
(2) eliminate/reduce use of inks containing heavy metals such as lead, mercury, and cadmium;
(3) modify Corporate Identity Guidelines for the company logo to allow for the use of recycled paper on office stationary and packaging materials;
(4) apply Society of Plastics Industry and American Paper Institute recycling symbols for appropriate packaging;
(5) require increased use of recycled and recyclable products;
(6) promote sale of single package, multi-product medical kits and reusable shipping containers;
(7) minimize using chlorine-bleached papers and paperboard in packaging;
(8) develop programs to encourage its suppliers to adopt similar packaging reduction packages.

Apart from the specific initiatives, Baxter has also established awards for recognizing individual as well as team performance. Three categories of annual awards have been instituted. First, there are awards for environmental managers in facilities and divisions. Facility-level awards are of two kinds: award for the best all-around program and for the best pollution-prevention initiative. Analogously, there are two division-level awards: all-around best program and pollution prevention. Second, the Erwin Awards, instituted in the memory of Dr. Lewis Erwin (who chaired Baxter's Task Force on environmentally sound packaging) are given to teams that have demonstrated excellence in packaging initiatives. Third, the Merit Awards, are presented to non-environmental managers that have significantly contributed to environmental efforts at the facility, division, company, or community levels.

Baxter has also pioneered corporate environmental accounting by calculating the business impact of its environmental initiatives. This is an important step for retaining the support of internal constituents that may perceive environmental programs as reducing profits. As shown in appendix 3.2, the 1995 Environmental Financial Statement suggests that the total costs of environmental programs for 1995 was $25.2 million while savings were $15.2 million. However, recurring yearly savings due to past environmental initiatives were $72.2 million. As a result, total savings to date due to environmental initiatives are $87.4 million. Thus, green accounting seeks to demonstrate the financial viability of environmental projects. Baxter's pioneering role in green accounting has been recognized by a variety of experts including Stephan Schmidheiny, the Chairperson of the World Business Council for Sustainable Development (Schmidheiny and Zorraquin 1996), and a major player in the 1992 Rio Summit.

Is Baxter's commitment to a state-of-the-art program financially viable and therefore sustainable in the long run? Baxter has very efficiently managed its environmental initiatives. Consider Baxter's expenditure on environmental programs as a percentage of sales. Baxter has succeeded in this as well, as it devoted only 0.3 percent of its 1995 sales to environmental programs. This was considerably less than the 2 percent figure reported by the 1991 survey of 220 companies done by of Booz, Allen and Hamilton and the 1995 study of Cooper and Lybrand that focused on firms with sales greater than $5 billion (Baxter 1995a).

Organizational structure

Since 1993, Baxter has steadily consolidated its environmental programs. Its organizational structure from 1992 to 1996 is summarized below:

Level I: Board of Directors; Public Policy Committee of the Board
Level II: Environmental Review Board (ERB); European Environmental Board
Level III: Corporate Environmental Affairs (CEA); Corporate Environmental Engineering (CEE)[2]
Level IV: Division Environmental Managers (DEM)
Level V: Facility Environmental Managers (FEM)

The Public Policy Committee of the Board of Directors is the highest decision-making body on environmental issues. This committee has two functions. First, it annually reviews Baxter's environmental performance. Second, it reviews the environmental plans of Baxter's operating units. Since its members are not employed by Baxter, this committee also provides an external oversight over the environmental programs.

The ERB is the highest internal organization on environmental issues. The ERB oversees all environmental initiatives, company wide as well those pertaining to divisions only. In early 1996, the ERB had eleven members: nine representing senior managers from different functional areas, and two DEMs. To ensure transparency in its decision making, all Baxter's environmental personnel are invited to attend the ERB's meet-

[2] In the new Baxter International Inc. (post 1996), CEE has been merged with CEA. Further, the Health and Safety Function has also been merged with the environmental function. As a result, William Blackburn in now Vice-President of Environmental, Health, and Safety (EHS). As I discuss subsequently, in Lilly, EHS constitutes a single corporate group.

ings which are held once every two months; and some of them actually do so.

The ERB is expected to champion environmental programs and provide top management support to them. As Loucks noted in his address to the conference on "Our Environment: A Healthcare Commitment," held in Arlington, Virginia, on March 10, 1992:

> My final and *most important recommendation is to provide strong top-management support* for your environmental program. Do this visibly, repeatedly, and sincerely. Serve as the main salesperson for the program to middle management. . . . Underscore the need for your managers to allocate appropriate resources. Recognize excellence in environmental performance. . . . Without this support, without this leadership, your environmental programs cannot succeed. *To say it starts at the top is really a cliché. It starts there or it doesn't start at all.* (Baxter 1992a: 10–11; italics mine)

Formed in 1992, the European Environmental Board oversees the implementation of environmental policy in Baxter's European Operations. Its status as independent of the ERB suggests that Baxter perceives environmental challenges in Europe as being significantly different from those in the US and other parts of the world. The head of Baxter's European operations is the *ex officio* chairperson of this committee. The head of Baxter's Corporate Environmental Affairs is also an *ex officio* member of this committee. In addition, as of March 1996, this committee had nine senior managers representing various European operations.

Corporate Environmental Affairs (CEA) coordinates environmental activities throughout the company. It also acts as a resource for divisions and facilities. It prepares the annual environmental performance report, organizes environmental audits, and manages legal issues relating to real estate transactions and superfund liabilities. In addition, it manages strategic initiatives such as estimating net savings from various environmental initiatives and evaluation of the ISO 14000 program (chapter 4). Corporate Engineering works closely with CEA on programs having engineering implications. It has managed key programs such as underground tank removal and toxic emission reduction (both chapter 4).

The DEMs are the main actors in initiating and implementing Baxter's environmental programs. Since they manage divisions' environmental budgets, they have significant influence in planning division-level initiatives. They also prepare their division's annual environmental performance report. This report is key to preparing Baxter's annual environmental performance report. DEM's support is often important for undertaking any company-wide environmental initiatives. The DEMs are also responsible for training the FEMs that report to them. The FEMs are responsible for managing all facility-level environmental programs, whether required by law or beyond-compliance. For example, they are expected to initiate and manage community-outreach programs. They are also required to train non-environmental managers on environmental issues.

In 1995, Baxter had 173 managers performing environmental functions. Since some of these managers had non-environmental responsibilities as well, there were 126 full time equivalents (FTE) environmental managers. Of these, ninety-six were at the facilities, twenty were at the divisions, and ten were at the corporate level. This suggests that Baxter's environmental programs are significantly decentralized, as only ten of the 126 FTEs are at the corporate level (Baxter 1995a).

Importantly, there has been a conscious effort to upgrade the quality of the FEMs; two strategies have been employed for this task. First, since training is often identified as a key factor in imparting skills, Baxter has established training norms: the FEMs of large manufacturing facilities are expected to receive a minimum of eighty hours of training per year and the FEMs of smaller facilities are supposed to receive sixty hours of training per year. This target was met in 1995 since, on average, the FEMs received seventy three hours of environmental training. The second strategy is to retain the FEMs, and thereby lowering FEM turnover. Since environmental laws are complex, it takes time for a new FEM to get accustomed to his or her role. As a result, the FEMs are in a position to initiate and manage beyond-compliance programs only after the adjustment period. The DEMs and corporate groups have made significant efforts to attract and retain managers committed to the environmental function. This has often involved significant personal commitment to train and support the new FEMs.

Eli Lilly: an overview

Founded in 1876, Eli Lilly and Company is an Indianapolis-based multinational corporation that develops, manufactures, and markets pharmaceutical products. As shown in table 3.2 below, its 1995 net sales stood at $6.7 billion, and its research and development (R&D) expenditure at $1 billion.

Lilly focuses on five disease categories: central nervous system, endocrine, infectious, cancer, and cardiovascular. Such focus is important since researching and developing new drugs has become extremely expensive. For reference, in 1995, the average cost to discover and develop a new drug was about $350 million and the average length of time to commercialize a drug was in the range of ten to twelve years (Lilly 1992a,1995b). Such cost pressures have also led to a spate of mergers and

Table 3.2. *Eli Lilly and Company: Financial highlights (in millions of dollars)*

	1995	1994	Change
Net sales	6,764	5,711	18%
Income from continuing operations	1,307	1,185	10%
Return on shareholder's equity	42.5%	25.9%	
Research and development expenditure	1,042	839	24%
Research and development expenditure as a percentage of net sales	15.4%	14.7%	

Source: Lilly (1997a).

acquisitions in the pharmaceutical industry across continents, as no single national market can defray the expensive R&D costs. Instead of merging with other drug firms to become a major player in a wide-range of pharmaceuticals, Eli Lilly has chosen to focus its R&D efforts on five disease categories. These categories represent its core-competencies (Prahalad and Hamel 1990) in which it has the critical R&D capabilities. A recent study commissioned by Lilly also suggested that size of firm *per se* is unimportant, once the firm has critical R&D capabilities in a given disease category (Lilly 1994a).

Let me elaborate with an example. Lilly has chosen to focus on diabetes, a disease in the endocrine category. Historically, it has been a leader in diabetes treatment, having developed and marketed the first insulin product in 1923. Many diabetics find that this disease disrupts their normal daily schedules: they are required to check their blood glucose several times a day with cumbersome instruments; they have to take insulin at least thirty minutes before a meal. Lilly has responded to these unmet consumer needs by developing: (a) a user-friendly electronic device for administering accurate insulin dosages; and (b) an insulin analog, Humalog, that can be taken just before a meal. Thus, instead of providing partial treatments in a wide range of disease categories, Lilly has chosen to provide a wide range of treatments in a few disease categories. As a result, if Lilly has a presence in a disease category then it is one of a market leader. Such a culture of doing few things, but doing them well, is also reflected in its environmental programs.

Evolution of environmental programs

The story of the evolution of Lilly's environmental programs is less dramatic than that of Baxter's. Lilly's environmental programs have progressed slowly but steadily with only occasional lapses. Initially, environmental programs were driven by Lilly's business needs, specifically treating waste-water streams from its bulk manufacturing facilities. Lilly has three types of facilities:

(1) Bulk manufacturing facilities, where it manufactures antibiotics in bulk using three types of processes: chemical synthesis, antibiotic fermentation, and bio-synthesis. Chemical synthesis is intensive in solvent use. Bio-synthesis involves using micro-organisms with recombinant DNA. It does not require using solvents or waste-water in significant quantities.

    Antibiotic fermentation is intensive in waste-water use since it involves using living micro-organisms as inputs. Colonies of such micro-organisms are developed in large tanks that ferment solutions of corn and soybean meal, lard oil, and plant starches. The manufacturing process is extremely water intensive. Once the manufacturing is over, the fermented material and waste-water need to be disposed. If it is not done in an environmentally sound manner, this can pollute water streams and harm human and aquatic life.

(2) Fill and finish facilities, where the bulk antibiotics are put into capsules or other final product forms.

(3) Research and Development sites and Technology Centers.

Most of Lilly's early environmental initiatives took place in its bulk manufacturing sites at Tippecanoe and Clinton, both in Indiana. Manufacturing processes in these facilities involve both chemical synthesis and antibiotic fermentation.

Lilly has been a pioneer in developing and adopting waste-water treatment technology. As discussed in the case study on underground tanks (chapter 4), as early as 1952, Lilly hired Robert Lowe to install a state-of-the-art waste-water treatment plant in its Tippecanoe facility. In the 1960s, Lilly began supporting a study conducted by Professor James Gammon of DePauw University on aquatic life in the Wabash River. Since the Clinton and Tippecanoe facilities are located on the banks of the Wabash river, Lilly was interested in documenting the impact of effluent discharges from these facilities on the Biological Oxygen Demand (BOD). Specifically, it wanted to investigate whether the investments in waste-water treatment plants have lowered the BOD levels in the Wabash river. BOD is a critical indicator of hospitability of any waterbody for aquatic life: the lower the BOD levels, the higher is the oxygen available to support aquatic life. Gammon's study suggests that since 1975 the BOD levels in the Wabash river have steadily fallen and the Wabash river has become more hospitable for aquatic life.

The late 1970s marked the beginning of Lilly's crop nutrient program. As suggested earlier, the bulk facilities generate fermented wastes in significant quantities. For reference, between 1987 and 1993, they generated between thirty one and forty two million gallons of such waste annually (Lilly 1995a). Since this fermented waste is rich in nitrogen, it can be used as a plant nutrient. Though this is an environmentally safe way of disposing of this industrial waste, its excessive application can result in the leaching of soil and the formation of nitrates in the water table. By providing this fermented waste free of charge to farmers in the vicinity of the Tippecanoe and Clinton facilities, and suggesting ways for their scientific application, Lilly disposed of the waste in an environmentally safe manner and also earned the goodwill of the farmers.

Programs such as establishing waste-water treatment facilities, supporting the Wabash river study, and organizing the crop nutrient programs were handled predominantly at the facility level with little direction from the corporate office. The need to strengthen corporate environmental resources began to be felt towards the mid 1970s. This was primarily due to the complex nature of new laws and regulations, especially the Resource Conservation and Recovery Act (RCRA) of 1976, and the Comprehensive Environmental Response, Compensation, and Liability Act (CERCLA) of 1980. In 1975, to strengthen the environmental function, Bert Gorman was appointed as the head of Corporate Environmental Affairs. Over the next few years, Lilly increased the managerial strength of its environmental organization, both at the corporate and facility levels: from three full-time equivalent (FTE) managers in 1975, to about twenty-five FTEs by mid 1980s. This increase was due to both new hiring from outside of the company and transfers from other divisions within Lilly.

The year 1989 constitutes a landmark in the evolution of Lilly's environmental programs for many reasons. First, in this year, the Corporate Environmental Legal department conducted Lilly's first internal audit. This audit focused on Tippecanoe facility's compliance with RCRA. Since it revealed certain shortcomings, the senior management approved inducting more managers in the environmental organizations, both at the corporate and facility levels.

Second, in 1989, the Environmental Management Committee (EMC; see below) approved an investment of about $80 million to install Regenerative Thermal Oxidizers for reducing emission of methylene chloride from its Tippecanoe and Clinton facilities (chapter 4).

Third, in this year the Food and Drug Administration (FDA) inspected the Indianapolis facility and identified problems in quality control systems. In response, Lilly restructured its Quality Control/Quality Assurance (QC/QA) organization, with Robert Williams as its new head. Williams was a senior manager holding the rank of Vice-President.

Importantly, environmental affairs was added to Williams's portfolio signaling to the whole organization the increased importance of environmental issues. Don Brannon, from Research and Development, was moved in as the new Director of Corporate Environmental Affairs.

Fourth, in 1989, Lilly formally adopted the Chemical Manufactures Association's (CMA) Responsible Care program. Under this program, firms were expected to implement six Codes of Conduct, including community outreach programs. Since there was internal opposition to implementing community outreach programs, Lilly initially implemented only five of the six Codes; community outreach programs began only in 1993 (chapter 4).

Fifth, the underground storage tank removal program was also initiated in 1989. Under this plan, Lilly removed all underground tanks, and replaced them with above-ground tanks that had significant beyond-compliance features. The total cost of this program was about $100 million with the beyond-compliance expenditures amounting to $30–40 million (chapter 4).

Though firms often adopt environmental programs in response to new laws and regulations, they may also play important roles in influencing the evolution of such laws and regulations. On this count, 1990 was an eventful year in that it marked the onset of a visible role for Lilly in influencing proposed laws and regulations. Lilly's legal department identified problems in implementing the 1990 Clear Air Act Amendments. As a result, Lilly sued the Environmental Protection Agency. Though the two parties eventually reached an out-of-court settlement, this well-researched intervention significantly increased Lilly's credibility in the environmental-legal community as a serious player in air and water regulatory issues. In 1993, Indiana's Governor Bayh asked Lilly to nominate one of its managers to serve on the state task force for identifying appropriate mechanisms to fund Indiana's water and hazardous waste programs. In that year, Lilly's legal department also worked closely with the Indiana Department of Environmental Management on Title V Air Program Rules (Lilly 1993). For example, in 1992, Lilly's managers attended 205 off-site meetings, offered written comments on fifty-six proposed legislations/regulations, and gave testimony in six hearings. The comments and testimonies were given on a wide range of issues such as Voluntary Remediation, Agency Funding, Pollution Prevention, RCRA, SARA, and Asbestos Regulations (Lilly 1992b).

Organizational structures such as the CEA are by themselves insufficient to change behaviors of actors within a firm. Often, clear sets of rules – institutions – are required that lay out responsibilities and sanctions for non-performance. Thus, in 1991, to provide a clear communication to internal and external stakeholders on the objectives of Lilly's environmental programs, Eli Lilly issued its *Environmental Policy and Guidelines* (appendix 3.3). In these guidelines, Lilly reaffirmed its commitment to follow or exceed all applicable environmental rules and regulations. Since this document constitutes a public statement of Lilly's intent, it may also be viewed as a set of standards for evaluating Lilly's environmental programs.

Environmental Policy and Guidelines lays out the macro-objectives of Lilly's environmental programs. It is not a tool to monitor progress on the various individual programs. In the highest positivist tradition, Robert Williams believed that Lilly should be able to measure its compliance with *Environmental Policy and Guidelines*. He argued that though environmental audits were useful tools for this task, an annual report on the overall progress of environmental programs was required. Consequently, in 1992, Donald Brannon and Richard Lattimer, both of CEA, put together and presented Lilly's first *Environmental Annual Report*. Its introduction stated:

> The purpose of this report is twofold. The first is to highlight the achievements of the environmental program at Eli Lilly and Company. The second is to be forthright regarding the environmental challenges facing Lilly as an inspiration to employees to take the initiative to improve the environmental performance of the corporation. (1992b: 3)

In its earliest version, the *Environmental Performance Report* was a *de facto* consolidated audit report. However, over time, these reports have become more comprehensive and they now serve as a tool for communicating Lilly's progress on its environmental programs, whether or not it is reflected in environmental audits.

In 1991, two initiatives were implemented to create an environmental ethic within the company. First, a Waste Minimization Conference was organized involving engineers and scientists from the various divisions of Lilly. Participants presented project ideas to minimize waste generation. This conference has now become an annual feature. Further, in 1992, the first Global Environmental Managers conference was organized that involved participation of environmental managers from Lilly's subsidiaries all over the world. Don Brannon played a key role in organizing this conference.

Second, in 1991, an interesting program to "green" non-industrial aspects of Eli Lilly was initiated. The objective was to recycle material used in office environments. Four kinds of materials were identified: white paper, beverage cans, laser printer toner cartridges, and telephone books. Goals were also established for each category. For example, compared to the target of recycling 1.5 million beverage cans, about 1.3 million cans were recycled; compared to a target of recycling 3,100 tons of white paper, about 3,700 tons was recycled. Funds generated from this recycling program are donated to local charities (Lilly 1992b). These kinds of initiatives are important because they instill a kind of environmental ethic among employees who generally do not consider themselves a part of Lilly's environmental programs. In addition, such programs also enhance in-company visibility of environmental managers.

During 1990–1993, the number of FTE environmental managers (predominantly environmental lawyers and engineers), both at the facilities and in the corporate groups, increased from 218 to 340 (Lilly 1992b). This was due to two major factors. First, this was almost forced by the ever-increasing number and complexity of environmental laws and regulations at the federal, state, and local levels. Second, there was a significant expansion in Lilly's bulk manufacturing capacity. Since there are many regulatory issues involved in bulk manufacturing, this required significant increases in managerial strength of the environmental function.

The year 1993 marks the peak of the growth phase of the CEA. First, Robert Williams retired. As suggested before, Williams was a key actor in establishing Lilly's environmental organization and institutions. Since, his successors have been of the rank of a Director,[3] some managers interpret this as an indication of diminished clout of its environmental organization. Further, by 1996, as a result of gradual attrition, the FTE headcount of CEA came down to twenty-one from thirty-five in 1991. Such staff reductions reconfirm suspicion about the declining fortunes of the environmental function. However, this also represented the trend towards decentralization: empowering facilities by locating most of the environmental managers at plant sites.

However, the picture is not as bleak as it seems. In 1995, primarily due to the initiative of John Wilkins of CEA, Lilly began publishing an annual Environmental Health and Safety Report to share information on its environmental programs with external stakeholders. In 1996, Lilly's Tippecanoe facility volunteered to become a pilot site for the CMA's Management System Verification (MSV) program. Responsible Care has been criticized for its lack of external verifiability; firms self-certify that their facilities are implementing the various codes of Responsible Care (chapter 4). In response to this criticism, the CMA is conducting MSV on a pilot-basis, and Lilly's Tippecanoe facility is one of such pilots. Under

[3] The hierarchy is: Chief Executive Officer, Executive Vice-President, Vice-President, General Manager, Executive Director, and Director.

MSV, managers belonging to CMA member firms will audit the facility's progress on Responsible Care. Importantly, this is the first time that Lilly's environmental programs will be audited by external actors. As discussed in the case study on environmental audits (chapter 4), unlike Baxter, Lilly does not invite external actors for environmental audits. Hence, the first external audit, even though on a trial-basis, signifies an increased confidence within Lilly about the strength of its environmental programs.

Organizational structure

Though Lilly's environmental function is integrated with its health and safety functions, I will describe only the environmental organization. This organizational structure has the following components:

Level I: Public Policy Committee of the Board of Directors
Level II: Operations Committee
Level III: Environmental Management Committee
Level IV: Corporate Groups: Environmental Affairs, Legal, and Engineering; Environmental Coordinators at sites.

Lilly's Board of Directors has five committees: Audit, Compensation and Management Development, Public Policy, Finance, and Directors and Corporate Governance. The Public Policy Committee of the Board of Directors is the highest policymaking body on environmental issues. Since its members are not Lilly's employees, they provide external oversight over Lilly's environmental programs.

The Operations Committee is the highest internal body on environmental issues and reports to the Board of Directors. Its main function is to align Environmental Health and Safety (EHS) strategy with other strategic objectives of the company. All its members are Lilly employees and it is headed by Lilly's Chief Operating Officer.

The Environmental Management Committee (EMC) is the highest internal body that focuses *exclusively* on EHS issues. It monitors progress on the various environmental programs and approves the environmental budgets that have been proposed by the divisions. The EMC reports to the Operations Committee and its members consist of senior managers representing various functions. Corporate Environmental Affairs (CEA) is a corporate group that coordinates activities required to carry out Lilly's *Environmental Policy and Guidelines*. It reports to Vice-President of Bulk Manufacturing, who in turn reports to Vice-President of Manufacturing. Since most of CEA's managers have had prior experience at facilities, they blend technical and operational level experience with an expertise in regulatory issues. As a result, CEA serves as a credible resource that delivers managerial tools to managers at plant sites. CEA manages the agenda of the EMC and prepares Lilly's *Annual Environmental Report*. Its managers represent Lilly in the CMA. Importantly, CEA's managers monitor proceedings of the Indiana General Assembly that pertain to environmental issues, and lobby on behalf of Lilly. For example, John Wilkins of CEA is a registered lobbyist in the Indiana General Assembly. The CEA also advises Lilly's lobbyists in Washington, DC on environmental issues and, with the help of Corporate Environmental Legal department, prepares written comments on the proposed environmental laws. CEA also manages strategic initiatives such as the development and implementation of the Plant Site Environmental Compliance Listing System, environmental risk assessment, and environmental impact analyses (chapter 4).

Another key corporate group, Corporate Environmental Legal (CEL), is a part of Lilly's legal division. CEL represents Lilly in all legal matters such as superfund issues and real-estate transactions. It is the final authority in interpreting environmental regulations even though CEA may provide the initial advice. Along with CEA, CEL organizes and leads environmental audits of Lilly's sites, third-party contractors, and waste-treatment facilities.

Corporate engineering provides technical input to CEA, CEL, and the plant sites. Plant managers have considerable autonomy in deciding their environmental programs. They propose the environmental budgets with corporate groups often having only advisory powers. This decentralization of environmental function manifests in two other ways: (1) plant-site environmental coordinators report to facility managers; they do not even have a dotted-line relationship with corporate groups; and (2) in 1995, of the 334 FTE environmental managers, only thirty three belonged to Corporate Groups: twenty-two in CEA, eight in CEL, and four in CEE (Lilly 1995c).

To streamline its Environmental Health and Safety (EHS) function, Lilly is implementing a Three-Loop Operations Design Model (Lilly 1996a). EHS activities include a wide range of regulatory and performance-based programs requiring varying levels of expertise. To prevent duplicating functions across managers, an expertise-based division of labor is required that best matches the talents of environmental managers with the needs of the organization. The first-loop managers are expected to focus on activities designed to meet the operational requirements for implementing Lilly's *Environmental Policy and Guidelines*. As a result, most of such managers will be stationed at plant sites. The second-loop managers are expected to focus on tactical activities designed to improve existing management systems and technologies. The third-loop managers are expected to focus on strategic tasks. These managers are expected to possess detailed knowledge of regulatory and technological trends. Typically the second- and third-loop managers will belong to corporate groups.

Baxter and Lilly: a comparison of environmental organization

It is instructive to compare Baxter's and Lilly's environmental organizations. Their environmental organizations have many similarities. First, in both firms, committees representing outside directors are the highest decision-making bodies on environmental issues: the Public Affairs Committee for Baxter; the Public Policy Committee for Lilly. Thus, these firms provide for an external oversight over their environmental programs. However, both these firms have a committee of senior managers (EMC for Lilly and ERB for Baxter) that is the *de facto* highest decision-making body on environmental policies. These committees have representation from various functional areas; this again suggests that both Baxter and Lilly acknowledge the cross-functional requirements of environmental programs.

Second, both firms also have a corporate group (the CEA) to coordinate various environmental programs. Further, in both, the corporate engineering group is distinct from other corporate groups, thereby highlighting the relative autonomy of technical personnel from the "policy types."

However, their environmental organizations also differ on three counts. First, in Baxter, the CEA is a distinct functional area headed by a Vice President who reports to the General Counsel/ERB Chair. In contrast, in Lilly, Corporate Environmental Affairs is not a distinct functional area. It is headed by a Director who reports to the Vice President of Bulk Manufacturing. This suggests that the environmental function has more clout in Baxter than in Lilly. Also, within Lilly there is a potential conflict of interest as CEA organizes and conducts (along with Legal) audits of manufacturing facilities. However, some of these facilities also report to the Vice President of Bulk Manufacturing.

Second, in Baxter, the Corporate Environmental Legal (CEL) department is part of CEA. In Lilly, CEL is a part of Lilly's Law Department and not of Corporate Environmental Affairs. Since CEA managers also work on regulatory issues, there is a duplication in activities between CEA and CEL; this may constitute a source of conflict between these corporate groups.

Third, a key feature of Baxter's environmental organization is the important role played by the Divisional Environmental Managers (DEMs). Perhaps, this reflects the relative heterogeneous nature of Baxter's business operations with divisions representing distinct business portfolios. In contrast, since Lilly has a more focused business portfolio, it has little need to divisionalize to the extent that Baxter has, and, as a consequence, there is no intermediary layer of environmental managers between the corporate groups and the plant sites.

To conclude, this chapter briefly described the evolution of environmental programs in Baxter and Lilly. The next chapter (chapter 4) examines ten cases of environmental policymaking. Four of these cases (underground tanks, 33/50, ISO 14000, and Environmental Audits) are common to both firms, and one each is idiosyncratic to them (green products to Baxter and Responsible Care to Lilly). These cases are examined by employing power-based and leadership-based theories of firms that were presented in chapter 2.

Appendix 3.1

Baxter's Environmental Policy

Baxter's Environmental Policy was adopted in 1990. The policy is applicable to Baxter's operations worldwide.

1. ***Environmental Review Board.*** An Environmental Review Board (ERB) appointed by the Public Policy Committee of the Board of Directors of Baxter is responsible for overseeing implementation of environmental policy. The ERB will review and decide matters of environmental importance and will make an annual report to the Board of Directors.
2. ***Legal Compliance.*** Baxter will comply with all applicable environmental laws.
3. ***Risk Control.*** Baxter will not create unacceptable risks to the environment and will minimize risk to the company from previous, existing, and potential environmental conditions.
4. ***Waste Minimization.*** Baxter will aggressively pursue opportunities to minimize the quantity and degree of hazard of the waste that results from its operations. It will reduce toxic and chlorofluorocarbon air amission 60 percent by 1992 and 80 percent by 1996, from 1988 levels based on equivalent production.
5. ***Environmental Leadership.*** Baxter will work to become a leader in respect for the environment. It will establish and maintain an environmental program to be considered state-of-art among the Fortune 500 companies. Baxter will accomplish this goal by 1993 in the United States, Puerto Rico, and Canada, and by 1996 worldwide.
6. ***Environmental Coordinators and Managers.*** The manager of each manufacturing and distribution facility, and other division and group managers where appropriate, will appoint a qualified environmental representative to coordinate and manage the unit's environmental program. However, compliance with this Policy is not just the responsibility of these representatives; it is the responsibility of every employee and particularly every manager.
7. ***Training and Audit.*** Corporate environmental personnel, divisions, and facilities will provide coordinated, effective environmental training, awareness and audit programs as appropriate.
8. ***Unit Manager Responsibility.*** The manager of each unit of the company will assure that the following are accomplished by the unit wherever relevant:

   8-1 Determine the facts regarding generation and release of pollutants from its facilities and responsibly manage its affairs to minimize any adverse environmental impact.

8-2 Develop and implement its own environmental management program to comply with this Policy.

8-3 Select, design, build, and operate products, processes, and facilities in order to minimize the generation and discharge of waste and other adverse impacts on the environment.

8-4 Utilize control and recycling technology wherever scientifically and economically feasible to minimize the adverse impact on the environment.

For more details on the implementation of these policies, see Baxter's Environmental Manual.

Vernon R. Loucks, Jr.
Chairman of the Board
and Chief Executive Officer

Arthur F. Staubitz
Senior Vice President, Secretary.
General Counsel and
Chairman of the Environmental
Review Board

Source: (Baxter, 1995a).

Appendix 3.2 Baxter's environmental financial statement

Estimated environmental costs and savings worldwide ($ in millions)

Environmental costs	1996[1]	1995[1]	1994[1]
COSTS OF BASIC PROGRAM			
Corporate Environmental Affairs and Shared Multidivisional Costs	1.4	1.4	1.3
Auditors' and Attorneys' Fees	0.5	0.3	0.4
Corporate Environmental Engineering/ Facilities Engineering	0.6	0.7	0.8
Division/Regional/Facility Environmental Professionals and Programs	6.6	6.8	7.6
Packaging Professionals and Programs for Packaging Reductions	1.0	2.3	1.6
Pollution Controls – Operations and Maintenance	3.6	3.6	3.3
Pollution Controls – Depreciation	1.4	1.7	2.1
Total costs of basic program	15.1	16.8	17.1
REMEDIATION, WASTE, AND OTHER RESPONSE COSTS *(Proactive environmental action will minimize these costs)*			
Attorneys' Fees for Cleanup claims, NOVs	0.1	0.2	0.2
Settlements of Government Claims	0.1	0.0	0.0
Waste Disposal	3.1	2.5	2.4
Environmental Taxes for Packaging	0.8	0.3	0.0
Remediation/Cleanup – On-site	0.3	0.3	1.0
Remediation/Cleanup – Off-site	0.1	0.6	0.0
Total remediation, waste and other response costs	4.5	3.9	3.6
TOTAL ENVIRONMENTAL COSTS	19.6	20.7	20.7

Environmental Savings
(Income, savings and cost avoidance from 1996 initiatives)

Estimated environmental costs and savings worldwide ($ in millions) (cont.)

Environmental costs	1996[1]	1995[1]	1994[1]
Ozone-Depleting Substances Cost Reductions	0.6	0.5	1.9
Hazardous Waste – Disposal Cost Reductions	(0.2)	0.1	0.4
Hazardous Waste – Material Cost Reductions	(0.2)	0.1	0.2
Nonhazardous Waste – Disposal Cost Reductions	(0.9)	1.3	4.6
Nonhazardous Waste – Material Cost Reductions	1.3	(0.5)	4.6
Recycling Income	5.6	5.2	3.0
Energy Conservation – Cost Savings[2]	2.5	1.0	0.5
Packaging Cost Reductions	2.4	5.6	5.7
TOTAL 1996 ENVIRONMENTAL SAVINGS[3]	11.1	13.3	20.9
As a percentage of the costs of the basic program	74%	79%	122%
Summary of savings			
Total 1996 Environmental Savings[3]	11.1	13.3	20.9
Cost avoidance in 1996 from efforts initiated in prior years back to 1989[4]	93.5	81.3	73.7
TOTAL INCOME, SAVINGS AND COST AVOIDANCE IN REPORT YEAR	104.6	94.6	94.6

Notes:

[1] These amounts have been adjusted to reflect the spin-off of Allegiance operations.

[2] Cost savings for 1996 calculated from data. Amount for 1995 is conservatively estimated. Amount for 1994 is only for US energy-conservation projects involving lighting.

[3] Cost avoidance from initiatives completed in prior years is listed as a separate line item in this report.

[4] No carry-forward savings is included for 1995 and 1996 for reductions of those ODSs that are no longer available to Baxter. This year's report used a more accurate method of calculating this line item, taking into account current-year pricing and compounded growth in business activity. This produced a significant increase in these sums over that which would have resulted from the previous method.

Source: Baxter (1997b).

Appendix 3.3 Eli Lilly and Company

Environmental policy

Eli Lilly and Company's mission is to create and deliver superior health care solutions in order to provide customers around the world with optimal clinical and economic outcomes. This mission requires that the company operate all of its facilities worldwide in a manner that protects human health and the environment.

Environmental guidelines

Eli Lilly and Company intends to carry out its environmental policy with a spirit of continuous improvement in the following ways:

The company will design, construct, and operate its facilities in a manner that protects human health and minimizes the impact of its operations on the environment.

The company will encourage and expect each employee to be environmentally responsible.

The company will provide ongoing education and training to Lilly employees so that they will be prepared to deal with day-to-day environmental responsibilities as well as environmental emergencies.

The company will comply with or exceed all applicable laws and regulations. Where existing laws and regulations are not adequate, the company will adopt its own environmental quality standards.

The company will make environmental considerations a priority throughout the process of developing new products.

The company will encourage and promote waste minimization, the sustainable use of natural resources, recycling, energy efficiency, resource conservation, and resource recovery.

The company will communicate its commitment to environmental quality to Lilly employees, shareholders, vendors, customers, and the communities in which it operates.

The company will recognize and respond to the community's questions about its operations.

The company will actively participate with government agencies and other appropriate groups to ensure that the development and implementation of environmental policies, laws, regulations, and practices serve the public interest and are based on sound scientific judgment.

The company will regularly assess and report to management and the Board of Directors on the status of its compliance with this policy and with environmental laws and regulations.

Adapted from: Lily (1996b).

4 Baxter and Lilly: case studies

Firms have indeed come a long way in designing and implementing challenging environmental programs. From resisting and challenging environmental laws and regulations, they now are willing to go beyond-compliance. However, there exist variations within and among firms on adopting beyond-compliance policies, specifically Type 2 policies, and this book examines such variations in Eli Lilly and Baxter International. The previous chapters discussed the theoretical framework and laid out the broad institutional context in which Lilly and Baxter have developed their environmental programs. I argued that the neoclassical economic theory cannot adequately explain why firms such as Lilly and Baxter selectively adopt Type 2 policies. For this we need to "unpack" the firm and examine intra-firm dynamics. To do so, I suggested employing power-based and leadership-based theories in the new-institutionalist tradition and also drawing upon insights from institutional theory and stakeholder theory.

In assessing the desirability of an environmental project, managers often have multiple objectives and varying preferences. All objectives cannot be collapsed into a single dimension – maximizing quantifiable profits – that standard procedures of project appraisal require. The book focuses on the most widely accepted procedure by US-based firms – capital budgeting. Type 2 policies create a political space for discursive struggles, where intra-firm dynamics shape managerial perceptions on the long-term benefits and costs of such policies. Profits are no longer an objective criterion that is invariant across managers. If Type 2 policies are adopted, it is due to two types of processes – power-based or leadership-based. In the former, policy skeptics do not revise their opposition to Type 2 policies and as a result, Type 2 policies can only be implemented by top management imposition. There is a clear evidence of dissent. In leadership-based processes, policy supporters manage to persuade policy skeptics of the long-term benefits and hence are able to induce cooperation. Importantly, Type 2 policies may not be adopted due to the absence of power-based or leadership-based processes to affect intra-firm dynamics. That is, policy supporters either fail to capture the top-management and have it impose the policy or they fail to induce cooperation by convincing the policy skeptics of the long term benefits, although non-quantifiable, of a given policy. The conditions under which power-based and leadership-based processes are likely to succeed are examined in chapter 5.

This chapter employs power-based and leadership-based theories along with insights from institutional and stakeholder theories to examine Baxter's and Lilly's policy processes in ten cases. The data for these cases were gathered from many sources including interviews, attendance in meetings, and published and unpublished documents. To maintain the confidentiality of my sources, I do not identify my data source in any manner except when quoting from a published document.

There are six sections in this chapter. Section 1 examines policy dynamics on underground tanks and concludes that leadership-based theories best explain policy adoption by these firms. Section 2 focuses on the Toxic Release Inventory and the 33/50 programs that were again adopted by these firms due to leadership-based processes. Section 3 discusses the Chemical Manufacturers Associations' Responsible Care program in the context of Eli Lilly only. Lilly initially resisted this program due to power-based dynamics and subsequently adopted it due to leadership-based processes. Section 4 investigates policies on "green products" in the context of Baxter only and concludes that leadership-based processes best explain their adoption. Section 5 focuses on environmental audits. Lilly undertakes only internal audits while Baxter invites external auditors as well. Lilly's policy adoption is explained by leadership-based dynamics while Baxter's is explained by power-based dynamics. Section 6 examines why neither Lilly nor Baxter adopted ISO 14000 environmental management standards. It suggests that this was due to the absence of either power-based or leadership-based processes and then discusses factors external and internal to these firms that led to non-adoption of ISO 14000.

1 Underground storage tanks

Baxter and Lilly have committed millions of dollars to replacing single-walled underground storage tanks (USTs) located in their facilities without subjecting this policy to any kind of formal investment analysis. In formulating their UST programs, these firms were confronted with the following issues: should they remove USTs at all, if so, how many of them, over what period of time, with what kind of replacement storage tanks, and should this policy should be limited to US-based facilities only or extended all over the world. The variation in expenditure was significant for Lilly ranging from $30–40 million for the least ambitious (though still in compliance with laws) route to more than $100 million for the most ambitious beyond-compliance route. Lilly adopted the most ambitious route by removing all US-based USTs and installing expensive above-the-ground tanks. Baxter installed double-walled underground tanks which were 15–25 percent more expensive than the EPA acceptable versions. Further, even though not required by law, Baxter installed new tanks in its facilities all over the world. By 1993, Baxter had spent over $10 million on its UST program in North America alone. My conclusion is that leadership-based theories best explain Baxter's and Lilly's response to the UST program.

Why did UST removal become a key policy issue in the mid 1980s? Industrial facilities often use USTs to store liquids in the form of raw materials, work-in-progress, final outputs, and industrial wastes. In the mid 1980s, the EPA estimated that about 70–80 percent of the "regulated"[1] USTs were constructed of bare steel, most of them were over ten years of age with approximately one-third being twenty years of age or more. Steel tanks often corrode non-uniformly and eventually leak through small holes known as "holidays" (USEPA 1987). In 1988, the EPA estimated that as many as 25 percent of all USTs were leaking (USEPA 1988a, 1988b).

Leaking USTs can pollute soil and groundwater. In fact, USTs are reportedly the main cause of groundwater contamination in many states. This is alarming since about 50 percent of the US population depends on groundwater sources to meet its drinking water needs (USEPA 1988a). Though not all leaking USTs may pose public health hazards, many of them store chemicals containing carcinogenic substances. Inflammable pollutants such as petroleum from leaking USTs can also migrate through soil to nearby structures creating potential fire hazards.

If a UST leaks, clean-up costs can be substantial: in the range of $500,000–600,000 when both soil and groundwater have been contaminated and about $50,000–60,000 when only soil has been contaminated (Baxter 1995a). The issue gets complicated since, *ex ante*, it is difficult to

[1] The non-regulated USTs included: (1) farm and residential tanks holding 1,100 gallons or less of motor fuel; (2) tanks storing heating oil used on the premises where it is stored; (3) tanks on or above the floor of underground areas such as basements, tunnels, etc.; (4) septic tanks; (5) tanks for collecting waste or storm water; (6) flow-through process tanks; and (7) tanks holding less than 100 gallons (USEPA 1988a).

Table 4.1. *EPA's time plan for implementing underground storage tank (UST) guidelines*

Type of tank and piping	Install leak detection by	Install corrosion protection by	Install spill/overfill prevention by
Tanks and piping installed after December 1988	At the time of installation	At the time of installation	At the time of installation
Existing tanks installed			
Pre 1965 or unknown	December 1989	December 1998	December 1998
1965–1969	December 1990	December 1998	December 1998
1970–1974	December 1991	December 1998	December 1998
1975–1979	December 1992	December 1998	December 1998
1980–Dec. 1988	December 1993	December 1998	December 1998

Source: Adapted from US Environmental Protection Agency, *Musts for USTs* (1988b: 17).

establish whether a UST will leak at all and when it will leak. On the other hand, as the book discusses subsequently, replacing USTs with above-the-ground tanks or double-walled underground storage tanks (DWUSTs) is very expensive (Baxter 1995a).

In accordance with the 1984 Hazardous and Solid Waste Amendments (HSWA) to the Resource Conservation and Recovery Act (RCRA) of 1976, the US Congress mandated the EPA to create a comprehensive national-level UST regulatory program. USTs were defined as those tanks that have 10 percent or more of their volume, including piping, underground (USEPA 1988a). On May 7, 1985, the EPA issued an interim ban on installing new single-walled USTs for regulated products (petroleum products, wastes, and hazardous substances) and required upgrading existing USTs by installing leak detection, corrosion protection, and spill/overfill prevention systems. As shown in table 4.1, UST upgrading had a phase-in period depending on the age of the UST. For example, though leak detection systems were required to be installed for all existing USTs by December 1993, UST owners/operators could install corrosion protection and spill prevention systems by December 1998. Importantly, the new regulations exempted flow-through tanks, the ones whose output feeds into another tank. Such tanks store liquids temporarily since they are a part of the closed-loop production process. Consequently, leaks in such tanks can be easily identified in the next stage of the production loop to which such tanks are dedicated. About 50 percent of Lilly's USTs could be classified as flow-through and would therefore be exempt from the new regulations. However, Lilly chose not to take advantage of this loop-hole and removed all its USTs.

Baxter and Lilly: policy dynamics

Baxter and Lilly used USTs primarily to store chemicals. Baxter is the world's biggest supplier of intravenous fluids whose production processes are water-intensive. Lilly is a major manufacturer of antibiotics whose production processes, especially in the fermentation stage, are organic solvent intensive. Hence both firms required some sort of storage tanks for their normal business operations. These firms responded by requiring their facilities within the US (also Canada and Puerto Rico for Baxter; Lilly did not have USTs in Puerto Rico) to remove all USTs within a time frame that was more aggressive than what the EPA allowed. Baxter also committed itself to removing USTs in its facilities outside North America by 1995 (Baxter 1988). Policy skeptics were concerned that Baxter was investing far too many resources in the UST program in terms of installing expensive new tanks, and expanding this program to facilities outside the US where such tanks were not required. Further, given the huge financial strain of Baxter's then recent merger with American Hospital Supply Corporation (AHSC), the skeptics questioned the wisdom of such an expensive beyond-compliance program. Baxter, however, chose to adopt the more expensive route: install state-of-the-art tanks and extend this policy to its facilities all over the world.

Baxter implemented its UST program in two stages. In the first stage beginning April 1987, Baxter's Corporate Environmental Affairs Division hired Geraghty and Miller, an outside consulting firm, to assess the degree of soil and water contamination caused by existing USTs (Baxter 1985b). Previously, American Hospital Supply Corporation (that merged with Baxter in 1985) had hired Geraghty and Miller to inventorize their USTs. They reported that Baxter had about 180 USTs in North America. As summarized in table 4.2, a preliminary survey of 77 USTs revealed that 60 percent of these had contaminated subsoil or groundwater (Baxter 1991a: 23).

The UST program was launched soon after Baxter's merger with AHSC. As discussed in chapter 3, this merger raised a storm both within and outside Baxter, particularly regarding the top management high handedness. UST policy supporters therefore realized that, although they had set an ambitious agenda for themselves, they must not create an impression of championing yet another top–down program. Policy supporters therefore sought to consciously involve Baxter's employees in various aspects of the program including identification of the number and

Table 4.2. *Baxter's experience: 77 USTs at 32 sites*

	Percentage of sites	Average cost of clean-up/removal (in US$)
No contamination	41	18,500
Soil contamination only	34	61,000
Groundwater and soil contamination	25	262,000

Source: Adapted from *Baxter Environmental Review* (1990a: 23).

status of USTs. In one of its editions, Baxter's *Environmental Newsletter* noted that:

[R]ecently many of you received a questionnaire about the underground storage tanks (USTs) at your facility. This survey is the first step in the corporate UST program. The purpose of this program is to evaluate all of the USTs within the company. . . . The first phase of this program, the survey is providing information on the tank, the facility, the topography, and the groundwater. This information will be used to develop a prioritization scheme for all USTs within the company. Survey data, applicable regulations, and other information will be used to identify tanks that require a more thorough investigation. The final phase of the project will predict the environmental impact from any leaking USTs. . . . Because of high probability of USTs developing a leak, they pose potential liability. To minimize Travenol exposure, consider whether or not your facility really needs a UST or its current number of tanks. (1986: 13)

In the second stage, Baxter's Environmental Review Board (ERB) earmarked a budget, outlined a time schedule, and set up an internal organization headed by Ron Meissen of Corporate Facilities Engineering to remove USTs. Meissen had been with the company for more than fifteen years and enjoyed considerable credibility for his technical as well as inter-personal skills. Corporate Environmental Affairs and Facilities Engineering were required to regularly submit progress reports to the ERB. This signaled to the whole organization the high priority given by the top management to this program and bolstered the credibility of policy supporters on this issue.

High-priority programs may flounder if implemented hastily or if there is insufficient and irregular sharing of information. Since policy supporters (belonging primarily to Corporate Environmental Affairs and Engineering) were sensitive to this aspect, there was a constant flow of information to facility managers on issues such as progress on program implementation, penalties for violating EPA's guidelines, etc. For example, the *Baxter Environmental Newsletter* reported that:

Table 4.3. *Baxter's progress on UST removal program*

	Number of USTs	Number of USTs in US, Canada, Puerto Rico	Number of USTs in Other Locations
1980	240	200	40
1988	175	140	35
1990	60	30	30
1992	25	—	25
1994	6	—	6

Source: Baxter's *State of the Environmental Program* (1995a: 21).

The first enforcement order to correct alleged violations of underground storage tank rules was issued for not cathodically protecting a bare steel tank as required. Even though no leaks or contamination was involved, United Parcel Service was assessed a $9,000 fine and ordered to remove a 10,000 gallon gasoline tank from a truck terminal in Connecticut. (1988: 20)

As summarized in table 4.3, Baxter made steady progress removing all its USTs in North America by 1990 and in rest of world by 1995.

Lilly also adopted a comprehensive program for removing USTs from its US facilities. In 1986, after the prodding of policy supporters, Lilly's Environmental Management Committee (EMC) began focusing on removing USTs on a priority basis. The EMC met virtually every month in 1986 to develop a strategy on the new UST regulations. Since there was scant information within the company on the number of such tanks, William Wager of Environmental Engineering was asked to compile an inventory of USTs. Mr. Wager went about systematically identifying tanks in the US and non-US locations. He found that the USTs were of various ages and capacities. As shown in table 4.4, Lilly had about 300 USTs worldwide in 1986, of which 275 were in the US. Further, 97 percent of them were in two production facilities in Indiana – Clinton and Tippecanoe. As I discuss subsequently, this also created incentives for Clinton and Tippecanoe facility managers to lobby for removing USTs.

To implement the new UST regulations, Lilly's EMC constituted a task-force consisting of James Vangeloff, David Bowman, and James Lowes. This group was asked to design new storage tanks incorporating state-of-the-art technology, and to prepare budgetary estimates for their installation. All these managers had been trained as engineers and had been working for Lilly for at least ten years. Hence they had the technical expertise and organizational knowledge for this task. Importantly, since they had all worked in Lilly's manufacturing facilities, they had first-hand experience of manufacturing and maintenance operations. As a result,

Table 4.4. *USTs in Eli Lilly*

	1986	1991	1994
US facilities	295*	249	22
Clinton, IN	123	87	20
Tippecanoe, IN	155	155	—
Others	7	7	2
Non-US Facilities	*	24*	24*
TOTAL	295*	273	44*

Note:
* This is a partial list as many non-US facilities did not know or did not report these numbers.
Source: Estimated from table 36 of (Lilly 1995c) and page 24 Lilly (1992a).

they had credibility with the plant site managers and received active help from managers such as Roger Lruswick of the Tippecanoe facility.

Lilly's management asked this team to apply the principles of zero-based budgeting in determining whether a particular UST was required at all or whether it was in operation due to historical reasons. In the course of their work, this team found that about 10 percent of the USTs could be eliminated by combining streams, cutting back on back-up tanks, etc. (Vangeloff 1992). This is not surprising as many engineering projects are routinely over-designed; that is, excessive resources are devoted to performing a task. This represents both "organizational slack" (Cyert and March 1963; Liebenstein 1966) as well as path dependency in technological choices.

As indicated earlier, the new EPA regulations did not apply to flow-through USTs. More than 50 percent of USTs at Lilly could be classified as flow-through process tanks and therefore were exempt from EPA regulations. Hence, there was a debate within Lilly whether all USTs should be removed or only the ones which do not fall into the flow-through category.

There were three other major debates as well that had significant bearing on the overall cost of the UST program; first, should the new tanks be placed above the ground (the more expensive alternative) or underground? Note that the EPA regulations permitted double-walled underground tanks. Second, if placed above the ground, should they be horizontal (the more expensive route) or vertical? Third, should above the ground tanks be placed in an enclosed area? Managers at Lilly were thus faced with the following problem: the costs of adopting the most spartan route (and yet remain in compliance with the new regulations) versus the cost of the most ambitious route were significantly different; instead of spending more than $100 million on the most ambitious route, Lilly could spend as little as $30–40 million. The policy skeptics pushed for the least expensive route since the most expensive route required more than twice the amount of expenditure. They also cautioned the policy supporters against over-reacting to the new EPA regulations.

Lilly chose to remove all USTs, including the ones which could be classified as flow-through process tanks. The new tanks were above the ground but below the grade. The former means that tanks are not buried and soil is not placed against their outer walls. Below the grade refers to the elevation of the tanks; that they are below the ground or grade level. The new tanks were placed on platforms, which in turn were placed in a concrete bathtub-like structure.

Two rows of new tanks were placed in a modular fashion. Unlike other good designs such as buried double-walled tanks, or tanks in a vault filled with stones that relied on instrumentation, this design ensured that tanks could be visually inspected (Vangeloff 1992). The tanks were placed horizontally and not vertically. Though this required more space (and more expenditure), it reduced the risk of accidents and sabotage. I understand that one of the lead characters in a popular TV series of those times had a hobby of shooting at tall structures. Mangers were, therefore, very concerned that vertical tanks may tempt some people to actually try out what was being shown on the TV. On this count, horizontal tanks appeared safer.

The storage tanks were placed in enclosures heated to 40 F. This had two major environmental advantages. First, it protected tanks from rain. Tanks often overfill and rainwater can wash away spilled toxic liquids which pollute water bodies. Enclosures would prevent rainwater from washing off spilled toxins. Second, during winters (which are severe in Indiana, the location of 97 percent of the USTs), facility personnel are reluctant to go out in the open to carry out regular tank inspection and maintenance. Placing storage tanks in enclosed heated areas reduced such weather-inspired shirking although it increased costs by about 5–10 percent. Overall, the newly installed modular USTs cost almost twice as much as other EPA acceptable versions.

Analysis

Power-based explanations are not helpful in explaining policy adoption because the UST program did not cause any significant conflicts either in Lilly or in Baxter. This program was adopted consensually and I attribute this induced consensus to active intervention of key managers who succeeded in convincing policy skeptics of the long-term benefits of incorporating Type 2 features in the UST program. Thus, leadership-based explanations are most appropriate to explain Baxter's and Lilly's responses.

Why and how did the leadership mode succeed? Policy supporters focused on the hostile external environment as a key motivation for the firm to go ahead with the UST program. In face of external adversity, the policy skeptics were amenable to arguments of the policy supporters. Thus, policy supporters invoked the coercive character of the external factors to impact internal dynamics and usher in institutional change. Importantly, since policy implementation did not require significant organizational changes, thereby creating "losers," policy skeptics did not have incentives to resolutely oppose the program.

The 1980s witnessed a flurry of new environmental legislation and an increasingly aggressive posture of regulators and citizen groups toward polluting companies. To use the popular expression, the "adversary economy" (Chandler 1980; Marcus 1984; Vogel 1986) became even more adversarial. Media also portrayed leaking USTs as vivid symbols of corporate America's indifference to environmental concerns. Markets also began to demand that firms incorporate environmental concerns in their policies. In particular, firms faced difficulties in getting insurance coverage for damages caused by normal business operations. Since the 1970s, many general liability policies had been excluding coverage for pollution liability unless caused by unanticipated events. This, in part, was motivated by court cases where general liability was interpreted as covering gradual pollution and its impact on humans after long periods of time (NAPA 1986). This created a market for insurance coverage against gradual pollution and many insurance firms stepped in to tap this market. In 1985, in anticipation of the imminent Superfund legislation, the market for gradual pollution liability for chemical (non-petroleum) exposures collapsed; insurance premia skyrocketed by almost 200 percent and eleven of the fourteen major insurance suppliers withdrew from the market. If gradual pollution insurance were available, its liability was often limited to a maximum of only $10 million. Thus, buying insurance coverage for chemical USTs became problematic for many firms. UST policy supporters within Baxter and Lilly used this as an example of significant but not easily quantifiable costs of ignoring environmental concerns.

It is also instructive to examine other potential benefits that a UST program bestows on firms and how they impact managerial perceptions of the desirability of spending millions of dollars for incorporating Type 2 features in the program. Adding beyond-compliance features creates three types of benefits. First, for the public at large, they create benefits of cleaner water and non-contaminated soil. These benefits, having characteristics of public goods, are diffused across beneficiaries. Since firms cannot charge beneficiaries for these benefits, such projects do not increase firms' profits. Consequently, profit-maximizing firms have little incentive to supply such public goods.

Second, beyond-compliance features create goodwill for firms with regulators, local communities, citizen groups, etc. Importantly, this goodwill is an excludable private good accruing only to firms undertaking this policy and imparts benefits in many ways: firms get quicker approval for their environmental permits; regulators consult firms on new laws and regulations and incorporate their suggestions; and regulators treat minor environmental violations leniently. Since such benefits to firms cannot be quantified, managers assessing projects solely on the basis of quantifiable profits and costs are not interested in undertaking such projects.

Third, incorporating beyond-compliance features reduces future liabilities of firms in the event of any UST leaks. Lilly's managers argued for placing all new storage tanks in bathtub-like concrete containers so that the tanks could be visually inspected for leaks.

Though in both Baxter and Lilly, some managers were initially hesitant to commit huge funds for USTs that had no quantifiable benefits, policy supporters eventually won them over. In Lilly, I identify five such policy supporters: Neil Pettinga, Earl Herr, Crandle, Bert Gorman, and Daniel Carmichael. These managers had the expertise, networking capabilities, and personal credibility to influence corporate decision-making. Pettinga was the Executive Vice-President of Lilly with a background in Research and Development. Crandle, an engineer, was a Group Vice-President in charge of Lilly's manufacturing operations and reported to Pettinga. Herr, the successor to Crandle, was an Executive Vice-President of Lilly. Gorman, a biochemist by training with a Masters in Management, was the head of the Corporate Environmental Division; Carmichael is the Deputy General Counsel, Company Secretary, and a member of the Environmental Management Committee. Since these individuals represented a variety of skills, their advocacy of the UST program was credible across functional areas.

In Baxter, I identify four leaders who championed the UST program: G. Marshall Abbey, C.F. Kohlmeyer, Ray Murphy, and William Blackburn, all having high credibility and persuasive skills. Abbey was a Senior Vice-President, the General Counsel, and the head of the Environmental Review Board. He was also a member of Baxter's Management Committee, the highest corporate decision-making body. Kohlmeyer was the Vice-President of facilities engineering, the agency responsible for removing existing tanks and installing new ones. Murphy was the head of Corporate Environmental Affairs until the end of 1989, and the Assistant General Counsel. Blackburn was Baxter's Senior Counsel until the end of 1989, and subsequently the Vice-President of Environmental Affairs. All had been in the company for over fifteen years.

As discussed in chapter 3, Baxter's and Lilly's environmental policies have been significantly influenced by their history of problems with the EPA. Hence, these leaders emphasized that removing *all* USTs and incorporating Type 2 features had significant symbolic value for both the internal and the external constituents of these firms. They invoked their firms' previous experiences with the EPA and State Environmental Agencies to argue for proactive UST policies. Further, they argued that it was in the long-term business interest of the firm to remove USTs. Baxter's *State of the Environment* Report therefore observed that:

> Contamination by leaking underground storage tanks used to store petroleum or other hazardous substances poses an environmental risk and liability exposure that is likely to escalate with time. Given the high costs of dealing with contamination from underground tank leaks, the money being spent for tank removals now before conditions worsen is clearly money well spent. (1995a: 21)

Baxter treats every facility as a profit center; revenues and costs are separately calculated for every facility and capital expenditures are financed from facility budgets. As a result, facility managers have incentives to oppose expenditures which do not reduce their quantifiable costs. I therefore expect that these managers would have opposed the Type 2 features of the UST program. Policy supporters in Baxter anticipated this opposition and sought to overcome it by making UST removal a corporate-level program. Hence, expenditures for removing UST appeared non-rival to projects which a particular facility manager was promoting. For a facility manager, this program created concentrated benefits (even though non-quantifiable) and diffused costs. Hence it was a natural candidate for their support.

Further, policy supporters realized that since the favorable stance of the former policy skeptics may not last for ever, it was necessary to have an accelerated program for removing USTs. This was an astute political move as coalitions that push for a given policy are often temporary. *Baxter Environmental Review* therefore reminded facility environmental managers that:

> Financial reserve accounts have been established at the corporate level to fund the removal of all tanks. This fund will pay expenses for removal of the UST(s), environmental assessment of the subsurface conditions surrounding the tank(s), and clean-up of any contamination costs that may exist . . . *we do not expect the financial reserve accounts will be available to fund UST removal after 1990, so it is important to finalize UST removal and replacement now.* (1989: 21; italics mine)

In Eli Lilly, the policy supporters (belonging to Legal Affairs and Environmental Affairs) struck an alliance with facility managers. As discussed before, nearly 97 percent of Lilly's USTs were concentrated in its two biggest facilities – Clinton, Indiana and Tippecanoe, Indiana – and facility managers had incentives to replace all USTs with sophisticated above-the-ground tanks since it reduced risks associated with manufacturing operations. Thus, this coalition championed removing all existing USTs (both flow-through process as well as non-flow-through process) and replacing them with expensive modular tanks.

Lilly's leaders also emphasized that historically, Lilly has been on the cutting edge of the UST technology. As early as 1953, Lilly hired Robert Howe to construct a waste-treatment plant in its production facilities. Howe constructed a state-of-the-art treatment plant so that waste resulting from fermentation processes (required for producing an antibiotic – mycelia) could safely be put in landfills. Howe's pioneering efforts created a lot of credibility with regulators who often turned to him for advice. Hence, it was argued that the proposed UST program was in keeping with the tradition of being industry leaders in the UST technology.

Finally and importantly, in both firms, the UST program did not upset the status quo. Although task-specific groups were also set up, policy supporters consciously employed extant organizational structures to implement the new policy. Because it did not threaten their standing within the organization, policy skeptics had few incentives to persistently oppose this program.

To summarize, both Baxter and Lilly invested significant sums in Type 2 features of their UST program. However, such capital expenditures were not subjected to any formal investment appraisal procedure such as capital budgeting. Leadership-based theories best explain the adoption of this policy. Policy supporters did not impose this policy on policy skeptics. They succeeded in convincing policy skeptics of the long-term benefits of adopting the Type 2 features, particularly in light of the hostile external environment. Policy skeptics did not have incentives to steadfastly oppose such policies because they did not perceive themselves as losing from the UST program. Consequent to the persuasions of policy supporters, policy skeptics eventually changed their assessments of the long term benefits and costs, thereby favoring the incorporation of Type 2 features in the UST program.

2 The Toxic Release Inventory and the 33/50 programs

Baxter and Lilly have invested significant sums in Type 2 policies to reduce the releases of Toxic Release Inventory (TRI) chemicals without subjecting such policies to formal investment analysis. Lilly invested in excess of $80 million for installing state-of-the-art equipment for curbing TRI releases even though less-expensive equipment ($40–50 million) meeting compliance requirements was available. Baxter invested more than $10 million for reducing emissions of TRI chemicals and air toxics. Further, both Baxter and Lilly joined the 33/50 program. As indicated previously, only thirteen percent of target firms (and only 64 percent of the Top 600 companies) releasing TRI chemicals joined this program. My conclusion is that leadership-based theories are most appropriate to explain Baxter's and Lilly's response to the TRI and 33/50 programs.

A major challenge for environmental regulators has been to ensure that firms reduce their releases of toxic chemicals and share information with local communities on such releases. To meet this objective, the US Congress enacted the Emergency Planning and Community-Right-to-Know-Act (EPCRA) in 1986 and the EPA proposed the TRI and 33/50 programs to consolidate the achievements of this law. EPCRA was enacted primarily in response to the Bhopal disaster.[2] It requires state and local governments to develop emergency response plans for unanticipated releases of certain toxic chemicals. Section 313 of EPCRA requires firms with manufacturing facilities in the US, as well as certain federal government facilities, to submit annual reports (Form R) to the EPA on the quantities of Section 313 chemicals that they have released into the environment. It also requires the EPA to make this facility-specific emission data available to public. To this end, the EPA has developed a computerized database known as the Toxic Release Inventory (TRI).

Not surprisingly, the initial TRI reports created adverse publicity for firms. Many environmental groups published rankings of leading polluters – "the dirty dozens" – in their states and counties. Managers within firms were also shocked to realize that their facilities were releasing significant volumes of toxic chemicals. Consequently, firms faced pressures from external stakeholders as well as employees for reducing emissions of TRI chemicals.[3]

[2] On December 3, 1984, the Bhopal facility of Union Carbide released methyl isocyanide causing deaths of thousands of people living in the vicinity of the facility. This underlines the dangers of living in the vicinity of facilities using, manufacturing, or emitting toxic chemicals.

[3] The EPA has proposed increasing the number of substances that facilities have to report for TRI purposes. The CMA opposes the EPA's proposal. It believes that since the TRI database seeks to provide information to communities on the releases of toxic chemicals into the environment, the proposed additions to the TRI do not fit this profile. Further, the CMA contends that the goal of the TRI program was to facilitate pollution prevention and risk reduction. However, the EPA proposal introduces use reduction though material accounting. This was not a part of the 1987 TRI initiative. The CMA fears that material accounting data may give competitors access to confidential information having significant business implications (CMA 1996b, 1999).

The TRI reporting requirements were the first milestone leading to the 33/50 program. The second milestone was the 1990 PPA (Pollution Prevention Act). Under the PPA, the US Congress directed the EPA to address the lack of attention to source reduction in existing environmental laws. Source reduction can be achieved by modifying equipment and processes, designing new products, or changing raw materials in existing products. To create incentives for firms to reduce the releases of TRI chemicals and to implement the mandate of the PPA, the EPA proposed a program called 33/50 in February 1991. Under 33/50, the EPA encouraged firms with US-based manufacturing facilities to *voluntarily* commit to reducing their releases of seventeen chemicals by 33 percent by 1992 and 50 percent by 1995 with 1988 as the baseline.[4] The rationale for targeting these seventeen chemicals was that: (1) they have significant adverse impacts on human health and the environment; (2) they were used in large quantities by US-based manufacturing facilities; (3) their releases relative to their total usage were high, suggesting inefficiencies in the production processes; and (4) their usage as well as releases could be reduced by employing pollution-prevention technologies and practices (EPA 1994).

The first year of TRI reporting was 1987. Due to these new and extensive reporting requirements in which firms were relatively inexperienced, the 1987 data were not considered very reliable. As a result, the EPA chose 1988 as the baseline year for the 33/50 program (EPA 1994). The 1987 TRI report indicated that more than 16,000 US-based facilities released at least one of the seventeen 33/50 chemicals and a total of 1.48 billion pounds. These facilities were therefore identified as the target group for the 33/50 program. The EPA contacted nearly 10,000 facilities of which 1,300 (13 percent) agreed to participate in this program (Sarokin, 1999). This again indicates that an exclusive focus on external factors under-specifies our explanations of why firms adopt or do not adopt beyond-compliance policies.

The EPA Administrator wrote to the chief executives of firms owning these facilities inviting them to join the 33/50 program. By contacting the chief executives, the EPA sought to encourage a pollution-prevention philosophy in the highest echelons of Corporate America. Firms participating in 33/50 were promised certificates recognizing their participation and

[4] The seventeen chemicals identified under the TRI program were: benzene, cadmium and compounds, carbon tetrachloride, trichloromethane (chloroform), chromium and compounds, cyanides and compounds, lead and compounds, mercury and compounds, methyl ethyl ketone, methyl isobutyl ketone, methylene chloride, nickel and compounds, tetrachloroethylene, toluene, 1,1,1-trichloroethane, trichloroethylene, and xylenes (EPA 1994).

Table 4.5. *Major TRI and 33/50 program projects at Lilly*

Year of completion	Project	Cost in millions
1988	Upgrade thermal oxidizer at Tippecanoe	$2
1990	Install plume suppressors for fume incinerators at Tippecanoe	$6
1992	Install regenerative thermal oxidizers at Clinton	$37
1993	Install regenerative thermal oxidizers at Tippecanoe	$40
1993	Methylene chloride emissions reduction projects at Clinton	$1
1994	Install air emission control equipment for bulk pharma manufacturing at Tippecanoe	$9
1994	Connect C9 to the regenerative thermal oxidizer at Clinton	$1

Source: Lilly (1992b: 9).

public recognition through periodic EPA reports and press releases. The EPA promised to measure the success of this program only at the national level; not at a firm level. Thus if a firm did not meet its goals, it would not get adverse publicity. Firms also had the autonomy to focus their 33/50 goals on all/some of their facilities as well as all/some of the seventeen identified chemicals. Emphasis on aggregate releases instead of releases of individual chemicals and facilities provided flexibility to firms for developing 33/50 plans as a part of their overall corporate strategies.

Baxter and Lilly: response to TRI and 33/50 programs

As indicated in tables 4.5 and 4.6, both Baxter and Lilly took significant initiatives for reducing their releases of TRI chemicals and achieving the objectives of the 33/50 program (Currie 1995; Wilkins 1996). Both firms are charter members of the 33/50 program. Lilly has spent about $80 million in its Tippecanoe and Clinton facilities for reducing releases of TRI chemicals, especially the 33/50 chemicals. Baxter has spent about $10 million for reducing releases of TRI chemicals, air toxics, CFCs, and 33/50 chemicals.

Of the seventeen chemicals targeted under the 33/50 program, Lilly used toluene and methylene chloride in significant quantities; the latter accounting for over 70 percent of Lilly's releases (Lilly 1996b). Some scientists claim that methylene chloride is carcinogenic in animals by both oral and inhalation routes, and produces toxicity of the liver, kidneys, and the central nervous system. However, there is no conclusive evidence that

Table 4.6. *Major TRI and waste-minimization projects at Baxter*

Project description	Project location	Cost in million
Install ethylene oxide control devices	Anasco/CVG Mt. Home Cleveland Aibonito Jayuya Singapore Colombia	$4.0
Install ethylene oxide/CFC-12 recovery system	Uden	$0.4
Install capture and control system on part washing equipment	Irvine/CVG Mt. Home Anasco/CVG Anasco/Biotech Castelbar/Swinford	$2.3
Convert to water-based coagulant, and install catalytic converters	Johnson City Kingstree Malaysia Exam Malaysia Surgical	$1.7

Source: Baxter (1995a: 10), table 5.

methylene chloride produces cancers in humans. Nevertheless, Lilly adopted three strategies for reducing the releases of this chemical. First, given methylene chloride's propensity to evaporate quickly, Lilly introduced new manufacturing technologies and practices to improve containment within the facility. Second, to make methylene chloride harmless, it was to be centrally collected and incinerated. Third, Lilly focused on reducing methylene chloride's usage in its new products.

For the first task, Lilly's top management constituted a task force known within the company as the "fix-it" team. This team, led by Paul Wassell, sought to identify all possible leakpoints and valves, a tedious task given that manufacturing facilities have hundreds of miles of piping. At the same time, Richard Lattimer estimated the releases of methylene chloride by using sophisticated material-balancing equipment. This team had an estimate of the aggregate quantities of these leaks and the points where they actually took place. To correct this gigantic problem, the team first experimented with many engineering and system innovations on a pilot basis in the Clinton facility. It succeeded in reducing emissions by 95 percent in one of the pilot sites. This innovation was then replicated in other sites within the Clinton facility, and, subsequently, extended to other Lilly facilities. Apart from the fix-it team, the role of the "SARA group" (in-company task force to implement Superfund Amendments and Reauthorization Act) was critical in organizing Lilly's attempt to reduce the releases of TRI chemicals. John Wilkins and Richard Lattimer, both of Corporate Environmental Affairs, played critical roles in organizing monthly meetings for tracking progress and intra-Lilly pollution-prevention conferences. As a result, there was a fair degree of employee involvement in programs for reducing emissions of TRI chemicals.

Methylene chloride continued to be emitted, albeit in smaller quantities. To reduce these emissions further, Lilly invested about $80 million in its two facilities at Clinton and Tippecanoe for installing state-of-the-art Regenerative Thermal Oxidizers (RTOs) or Fume Incinerators. All fumes from point sources, especially those containing volatile organic chemicals such as toluene and methylene chloride, were centrally collected and incinerated in these RTOs. Since installing the RTOs involved significant Type 2 expenditures that were not subjected to capital budgeting, they were initially opposed. However, policy supporters, especially Edward Smithwick, Vice-President of Bulk Manufacturing, and Richard Eisenberg, Executive Director of Corporate Environmental Affairs, emphasized the need for proactive measures, even though their benefits were non-quantifiable. The proposal for installing the RTOs was initiated in 1988. Due to initial opposition, the RTOs were finally installed in the Clinton facility in 1992 and in the Tippecanoe facility in 1993. Lilly also attempted source reduction by identifying non-toxic substitutes of methylene chloride. This involved employing a new management system known as NPERT (New Product Environmental Requirements Tracking). NPERT requires identifying environmental challenges in the early stages of process development of new drugs. Lilly employed NPERT in developing Raloxifene, a drug which can potentially treat and prevent osteoporosis. The Raloxifene team successfully eliminated methylene chloride and replaced it with a less volatile solvent that was easier to contain within the plant simply by altering the process chemistry. This team also eliminated/reduced other solvents (specifically polyphosphoric acid) from the manufacturing process resulting in a 74 percent reduction in total solvent use compared to the previous process. As shown in table 4.7, consequent to the efforts of the fix-it team, the new RTOs, and the NPERT process, Lilly reduced its releases of 33/50 chemicals by 81 percent between 1988 and 1995, well exceeding EPA's target of 50 percent reduction. The releases of methylene chloride, the key 33/50 chemical,

Table 4.7. *Lilly's performance on the 33/50 program*

	Total releases (in pounds)	Index versus baseline year 1988
1988	575,000	100
1991	450,000	69
1992	780,000	136 (EPA target: 67)*
1993	500,000	87
1994	100,000	17
1995	110,000	19 (EPA target: 50)

Note:
* This increase in releases was due to launching of new products requiring 33/50 chemicals for their manufacturing.
Source: Estimated from Lilly (1992b, 1995c).

were reduced by 77 percent by 1995 compared to 1988 levels. This team's efforts and the NPERT process were awarded the 1995 Indiana Governor's Award for Excellence in Pollution Prevention.

Baxter also adopted beyond-compliance policies in response to TRI. As reference, Baxter used at least eight of the seventeen 33/50 chemicals. These were: chloroform; chromium and chromium compounds; methylene chloride; methyl ethyl ketone; nickel and nickel compounds; toluene; 1,1,1,-trichloroethane; and trichloroethylene. Of these eight chemicals methyl ethyl ketone and 1,1,1,-trichloroethane were used in significant quantities (Baxter 1991a). As discussed previously, the TRI database has given external stakeholders access to information on facility-wise releases of toxic chemicals. Consequently, during 1989–1991 Baxter faced considerable adverse publicity in many states and counties where its facilities were located. This prompted some managers to suggest aggressive programs for proactively reducing the releases of TRI chemicals. A technical team within Baxter headed by Vice-President, Technical, Charles Kohlmeyer was established for evaluating the technical feasibility of reducing emissions of key TRI chemicals. A number of task-forces were set up to focus on individual chemicals. For example, Kohlmeyer himself chaired the task force on ethylene chloride and methanol and Rob Currie chaired the task force on trichloroethane. Consequently, Baxter has spent over $10 million on projects for reducing the releases of TRI and other chemicals.

These proactive steps for reducing the releases of TRI chemicals had significant *ex post* payoffs. In 1991, when the EPA invited Baxter to join the 33/50 program, Baxter already had on-going programs for meeting most of the 33/50 goals. On May 14, 1991, Vernon Loucks, Baxter's chief

Table 4.8. *Baxter's performance on the 33/50 program*

	Total releases (in pounds)	Index versus baseline year 1988
1988	873,000	100
1992	86,000	10 (EPA target: 67)
1994	42,000	5
1995	37,000	4 (EPA target: 50)

Source: Baxter (1996a).

executive, wrote to the EPA on Baxter's commitment to joining the 33/50 program. In his letter, Loucks committed to even higher standards than were required by the 33/50 program. Instead of a 50 percent reduction in releases by 1995, Baxter set a target of 80 percent reductions. In addition, Baxter established an 80 percent reduction goal based on equivalent product output for CFCs and 189 other chemicals covered by the Clean Air Amendments (Baxter 1991b). Importantly, Baxter set these *minimum* goals for facilities all over the world; facilities and divisions could set even more aggressive goals for themselves.

Baxter followed a three-stage process to meet the 33/50 program goals. It accorded source reduction the highest priority. If source reduction was infeasible, recycling was to be considered. If recycling was also infeasible, treating chemicals (and converting them to harmless compounds) was to be the route for meeting the 33/50 program goals. However, treatment was to be considered only as an interim measure; corporate research and development teams continued efforts to identify alternative processes or materials.

As shown in table 4.8, Baxter reduced its releases of the 33/50 chemicals by 96 percent: from 900,000 pounds in 1988 to about 37,000 pounds in 1995. This was a significant achievement because during this time period company-wide production increased by 20 percent.

Some of the key projects for implementing the 33/50 program are described below:

(1) Bentley Laboratories in Irvine, California, develops, manufactures, and markets disposable medical devices used for blood oxygenation during cardiopulmonary surgery. This facility reduced the emissions of 1,1,1-trichloroethane in two stages. In stage one, emissions were reduced by over 80 percent by 1990 by modifying process leach tanks and instituting operational changes. Initially, this facility resisted undertaking capital expenditure of about $50,000 for this task. However, persistence and persuasion by some managers, especially

Rob Currie, resulted in implementation of these initiatives. In stage two, by adopting a newly developed proprietary technology, which eliminated using 1,1,1-trichloroethane, these emissions were reduced to zero by 1991.

(2) The V. Mueller division manufactures a complete line of general and specialty surgical instruments, a line of endoscopic products for laparoscopic surgery, and the Genesis Container system. Mueller's St. Louis, Missouri facility totally eliminated the releases of 1,1,1-trichloroethane (8,000 pounds per annum in 1988) by installing an aqueous-based cleaning system.

(3) The Mannford facility in Oklahoma which manufactures a broad line of surgical and medical products such as disposable/reusable gowns, disposable drapes, and surgical packs, reduced its emission of methyl ethyl ketone by 60 percent between 1988 and 1994. This was achieved by modifying manufacturing process and reducing fugitive losses.

Although not required under the TRI or the 33/50 programs, Baxter's international facilities also significantly reduced their toxic emissions. Some examples:

(1) The Euromedical facility in Malaysia reduced methylene chloride emissions by over 100,000 pounds with 1988 as the baseline. This was achieved through changing material handling and modifying equipment.

(2) Methyl ethyl ketone emissions in the various maquiladora facilities located in Mexico near the US border were reduced by over 90,000 pounds with 1988 as the baseline by operational changes and tank modifications.

Analysis

TRI and 33/50 programs required Baxter and Lilly to spend millions of dollars in installing new equipment. Similar to intra-firm dynamics for investing in underground tanks, power-based explanations are inadequate since there was little dissent or bickering over the policies on TRI and 33/50. The reason is that these programs did not create "losers" who would have incentives to resolutely oppose them. There was indeed some initial opposition. However, after initial stalling, policy skeptics were persuaded to revise their assessments of the long-term benefits and costs of Type 2 expenditures on these programs. The relentless campaign by policy supporters played a significant role. In particular, they emphasized the critical need to remain in the good books of the EPA, an important external institution, particularly by adopting the 33/50 program. Thus, leadership-based explanations best explain the policy responses of these firms.

How did policy supporters convince policy skeptics? As indicated previously, the initial TRI reports generated adverse publicity for many firms. In 1988, Lilly was identified as one of the top ten emitters of TRI chemicals in the state of Indiana. Baxter was listed as a leading polluter in many states and counties (appendix 5.1). Further, *even when some firms reduced their releases of TRI chemicals, they continued to be identified as emitters of toxic materials*. For example, The *Arkansas Gazette*, in its eye-catching caption "Toxin level drops by half in 1 year, study says," reported that:

> The latest figures from the federal Toxics Release Inventory show that Arkansas industries discharge less than half as many toxic chemicals in 1990 as in 1989.... Another example of dramatic reduction was Baxter Healthcare Corp. which reduced its total toxic emissions from 609,600 pounds in 1989 to 432,308 pounds in 1990. (1992: Arkansas Page)

Even complementary news items, such as the aforementioned, reinforce an image that firms continue to be significant emitters of toxic chemicals. For policy supporters, the message was clear: firms need to carefully protect their reputations because negative images tend to be sticky. Clearly, adverse media coverage created pressures on both Baxter and Lilly from their external stakeholders as well as employees for reducing releases of TRI chemicals.

Policy supporters further argued that the TRI program was a strategic step by the EPA to prepare the ground for stringent regulations. They suggested that the 33/50 program was a prototype for new laws if "voluntary" compliance from industry was not forthcoming. They highlighted the letter written by the EPA's Administrator, William K. Reilly to Vernon Loucks, the Chief Executive Officer of Baxter, noting that:

> The American public has made clear that they expect nothing less than dramatic reductions in toxic chemicals releases. The challenge before EPA and industrial leaders is this: how do we bring such reductions about? One way is by the conventional command-and-control option which has been the Agency's mainstay for the past twenty years. But I believe there is another, more fruitful path that we can follow which is faster, and *without the detailed direction which is likely to be demanded by the public if voluntary efforts are not fruitful* (Baxter 1991a; italics mine).

For policy supporters, Reilly's message was very clear: either industry voluntarily commits to such reduction, or citizens will demand that the government intervene. Such interventions, typically in the command-control mode, will leave little autonomy for firms on issues such as which chemicals to reduce, which technology to employ, and the time frame of implementation. Assuming that Riley's threat was credible, what incentives did individual firms have to actually join the 33/50 program or to voluntarily reduce the release of TRI chemicals? Why could individual firms not free-ride on the efforts of other firms? How did the institutional design of the 33/50 program and the TRI reporting requirements mitigate collective action dilemmas?

The TRI reporting requirements and the 33/50 program created two categories of benefits. First, citizens benefitted from having a cleaner environment. Since these benefits are non-excludable, firms producing such benefits cannot force citizens to pay for them. As a result, firms have little incentive to provide these benefits. Second, these programs create goodwill benefits for firms among regulators and citizens. These programs also raise the morale of employees. Such goodwill benefits accruing to firms are excludable since it is possible to identify firms that have invested in reducing the releases of TRI chemicals or have joined the 33/50 program. Citizen groups and journalists often monitor firms' performance on reducing TRI emissions, and advertise them. The EPA also enhances the excludability of such benefits by regular media reports on 33/50 success stories or trends in releases of TRI chemicals.

What were the strategies adopted by policy supporters to convince their skeptical colleagues of the benefits of such Type 2 programs? As discussed previously, the first TRI report was sent to the EPA in July 1988. In the process of quantifying releases of TRI chemicals, policy supporters in Lilly realized the potential for bad press once this information became public knowledge. As early as 1988, they suggested that Lilly should be prepared to invest millions of dollars to proactively reduce its emissions of methylene chloride. At first, there was resistance to this proposal since its benefits were not quantifiable. Skeptical managers argued that since Lilly adheres to existing emission laws, there was no need for further reductions at a substantial cost. However, the supporters pointed out that consequent to the TRI reports, Lilly had been identified as one of Indiana's leading emitters of TRI chemicals. The Clinton and Tippecanoe facilities had been ranked fourth and fifth nationally as emitters of methylene chloride. In addition, it is necessary for Lilly to remain on good terms with the EPA and proactive reductions in emissions of TRI chemicals would facilitate this.

Edward Smithwick, Vice President, Bulk Manufacturing, and Richard Eisenberg, Executive Director of Corporate Environmental Affairs, in particular, were key figures in persuading their skeptical colleagues of the long-term benefits, even though not quantifiable, of proactively reducing TRI emissions. The advocacy by Smithwick was important since the expenditures on reducing TRI emissions were borne by facilities. Since every facility is a profit center, one expects facility managers and their organizational superiors (such as Smithwick) not to support programs that reduce quantifiable profits. Once the TRI chemical reduction program was in place, 33/50 targets could be achieved easily. Consequently, in 1991, when the EPA invited Lilly to join the 33/50 program, policy supporters met with only marginal opposition. The skeptics saw little value of joining yet another EPA-sponsored program. However, policy supporters pointed out that since Eisenberg served on the Federal Advisory Committee on the Clean Air Act, it was difficult for Lilly not to join the 33/50 program. Further, they pointed out that the 33/50 program would require only minor changes in the internal organization. This of course reassured the skeptics that the organizational status-quo would not be upset which may have inflicted "losses" on them.

Baxter also invested in excess of $10 million for reducing its releases of the TRI and other toxic chemicals. Policy supporters adopted an interesting strategy by making this project a part of Baxter's overall objective of establishing a state-of-the-art environmental program. Since the policy skeptics could not exercise a line-item veto, and they were in favor of other items of the state-of-the-art program, they had few incentives to derail the whole program.

As discussed in chapter 3, Baxter's environmental program faced rough times in the mid 1980s primarily resulting from Baxter's merger with the American Hospital Supply Corporation. In 1989, some of Baxter's senior managers realized the long-term costs of having reactive environmental policies. This coalition of senior managers, specifically, William Blackburn, Charles Kohlmeyer, and G. Marshall Abbey (and, subsequently, Arthur Staubitz), argued for establishing a state-of-the-art environmental program within Baxter. They eventually managed to convince Vernon Loucks, Baxter's Chief Executive, of its long-term payoff. Vernon Louck had presided over the controversial merger of Baxter with AHSC and was therefore inclined to take necessary steps that would reduce the potential flak on environmental issues. In the process of operationalizing this ambitious mandate, policy supporters included programs for reducing the releases of TRI chemicals. They very astutely linked these programs to reductions in the emissions of CFCs, chemicals that had been identified as main causes of the Ozone Hole and whose use and production had been regulated under the Montreal Protocol of 1987. Further, the TRI reduction program was subsumed under a broader program for waste reduction and pollution prevention. These were very clever political strategies where by TRI and 33/50 programs were bundled with popular and established policies.

Nevertheless, there was opposition from facility managers on two counts. First, they were hesitant to commit to ambitious and still undefined state-of-the-art goals. Second, since facilities were profit centers, they bore the costs of such programs. They were therefore reluctant to bear costs for projects that imparted meager excludable and quantifiable benefits to their facilities. Further, they feared intrusion by the Corporate department in their environmental policies under the garb of these programs.

Policy supporters read this opposition carefully. They decided to induce acceptance by addressing the concerns of policy skeptics through persuasion and not imposition. First, they finalized the program goals only after actively consulting with the facility and divisional managers. Thus, only after skeptical managers had been convinced of the technical feasibility and financial support for these programs did the policy supporters propose the program goals to the top management.

Second, policy supporters consciously decentralized the implementation of the various pollution prevention and waste minimization programs. This encouraged facility managers to take the lead in such programs, to identify solutions, and, importantly, to take the credit for their success. Thus, the decentralized structure aligned incentives of the corporate, divisional, and facility managers, thereby enthusing the hitherto skeptics to ensure that the programs were adopted.

The policy supporters regularly publicized the progress made by facilities and contributions of specific managers towards reducing the releases of TRI chemicals, 33/50 chemicals in particular (for example, Baxter, 1994a). Baxter's annual environmental conferences, often attended by about 200 managers, were extremely valuable in providing in-company visibility to high performers. Also, Baxter's bi-monthly in-house magazine *PACE*, circulated to all its 60,000 employees consciously highlighted the achievements of such programs. This created incentives for the facility managers to enthusiastically adopt this program since they did not want to be identified as Baxter's "bad guys." In his letter to the EPA, Rob Currie, Baxter's 33/50 coordinator noted that:

> Baxter has found that communication and awareness are key to implementing successful pollution prevention and waste minimization programs. As a result, several vehicles to foster communication are used to bolster facility programs . . . (Currie 1995)

To conclude, leadership-based theories best explain why and how Baxter and Lilly invested millions of dollars in reducing releases of TRI chemicals. The TRI and 33/50 programs were initially opposed on account of their significant financial commitments and their potential to bind facility managers to relatively undefined goals. Policy supporters did not trample upon this opposition, as power-based theories would predict.

Instead, they sought to overcome it by bundling these programs with other widely popular policies, and, importantly, by genuinely attempting to address the concerns of the facility managers. They also emphasized the need to remain in the good books of the external stakeholders, especially the EPA. Since the role of key managers was critical in inducing this consensus, the policy adoption is best understood by employing leadership-based theory.

3 Responsible Care[5]

This section examines Eli Lilly's response to the Chemical Manufacturers Association's Responsible Care program. Specifically, why did Lilly first resist and then enthusiastically adopt this program? Given the nature of its product portfolio (with one of the 200 facilities involved in manufacturing chemicals), Baxter has virtually no involvement with the CMA, and consequently with Responsible Care. Hence, this section focuses on Lilly only. My conclusion is that power-based theories best explain Lilly's initial reluctance, and leadership-based theories its subsequent enthusiastic adoption of Responsible Care.

Founded in 1872, the Chemical Manufacturers Association (CMA) is one of the oldest trade and industry associations in the US. Its 200 members together account for about 90 percent of US industrial capacity in basic chemicals (CMA 1996a). Responsible Care is CMA's flagship program for improving the chemical industry's performance in the areas of health, safety, and the environment. Versions of Responsible Care have been proposed by CMAs in other countries as well. The US CMA mandates that all its member adopt Responsible Care. Members, however, have the autonomy on the pace of its implementation, and, conceivably, members may not implement it at all.[6]

[5] This section draws on Prakash (1999a).

[6] Responsible Care was launched first in Canada and brought to the US and other countries only subsequently. The subsidiaries of US multinational enterprises (MNEs) such as Dow-Canada were early adopters of Responsible Care and played a significant role in establishing the legitimacy of this program in Canada within chemical firms. Many of Responsible Care's ideas were developed by Dow-Canada in response to the accident in its Sarina facility. Some individuals such as Dave Buzelli (Dow-Canada) and Robert Kennedy (Union Carbide) played key roles in popularizing Responsible Care in their firms, their US parent, and the Canadian/US CMA. It is said that during a review of Canadian operations, Robert Kennedy, Chief Executive Officer of Union Carbide, was informed of this initiative. He quickly gauged the potential benefits of having this as an industry-wide program in the US and passed this information on to the US CMA's Public Perception Committee. The committee bought into Kennedy's suggestion and consequently, the US CMA launched Responsible Care in 1988 *(Chemical and Engineering News* 1992).

Responsible Care was launched in the backdrop of economic success but stakeholder distrust of the US chemical industry. The industry's economic performance was impressive: it was ahead of most others in exports, research and development expenditures, and wages paid to its manufacturing workers (*Chemical and Engineering News* 1992). However, its credibility with regulators and other external stakeholders on environment, health, and safety issues was plummeting. A series of major chemical accidents, including the Bhopal disaster in 1984, reinforced a perception that the chemical industry cannot conduct its operations without harming human health and damaging the environment. Consequently, there were demands for governmental intervention. Industry leaders feared that such high levels of policy activism and the accompanying uncertainty would undermine investors' confidence in the long-term prospects of the chemical industry and hurt its stock prices. Further, they argued that the ever-increasing compliance requirements would divert scarce resources from research and development, and eventually hurt the industry's international competitiveness.

In 1983, one year prior to the Bhopal tragedy, the CMA had developed a statement of principles on how the chemical industry should conduct its business and relate to its stakeholders. This statement eventually became the basis for developing the Ten Guiding Principles of Responsible Care (see below). However, the scale of the Bhopal disaster altered the character of discourse on how the industry ought to be regulated; immediate and drastic actions were demanded. To introduce public accountability of its activities, the CMA proposed a voluntary program, Community Awareness and Emergency Response (CAER), in 1985. Eventually CAER became one of the Six Codes of Responsible Care (see below). Since many environmental groups saw CAER as a public relations gimmick, in late 1985, the CMA formed the Public Perception Committee composed of top industry executives. This was a precursor to the Executive Leadership Groups, an important feature of Responsible Care. The Public Perception Committee recommended that the CMA launch Responsible Care (*Chemical and Engineering News* 1992).

Responsible Care was launched in the US in 1988. Three categories of actors have subscribed to Responsible Care: (1) members of the CMA; (2) partner companies, particularly those in the transportation sector: railroads, trucking, and barge (these firms are not members of the CMA); (3) partner associations such as the state chemical industry councils and the national associations of firms that deal with chemicals.

This program has the following features (*Chemical and Engineering News* 1993b; CMA, 1996a,b):

(1) Ten *Guideline Principles* spelling out responsibilities of CMA member firms.
(2) Six *Codes of Conduct* that identify more than one hundred specific management practices. These codes seek to establish management systems in manufacturing, distribution, and transportation (details in appendix 4.2).
(3) A fifteen member *Public Advisory Panel* consisting of non-CMA members to guide Responsible Care. This panel is expected to sensitize the CMA to public concerns and give input on developing programs that better address these concerns.
(4) A *requirement* that all the member firms will adopt Responsible Care. Firms are not required to implement all the six codes immediately; they can chart their own time frame for implementation.
(5) Member firms must *annually evaluate* their progress on implementing the six codes of Responsible Care. This evaluation should be shared with the CMA.
(6) *Executive Leadership Groups* of senior industry representatives to periodically share their experiences on implementing Responsible Care and to identify areas requiring assistance from the CMA.

Critics are skeptical of the impact of Responsible Care on the culture and functioning of the chemical industry. They view it as an attempt by industry to preempt stringent legislations (Chatterjee and Finger 1994). Fred Miller of Friends of Earth views Responsible Care as "the velvet glove around the iron fist . . . industry has a heavy burden of proof to show that it is just not a PR gimmick" (*Chemical and Engineering News* 1992: 39). Further, many dispute the extent of change in the culture and practices of the chemical industry, and the speed with which they have been accomplished.

Further, since CMA membership is voluntary, a threat of expulsion for not implementing Responsible Care is not credible. This is also tacitly accepted by Jon Holzman, CMA's Vice President for Communication who maintains that the CMA does not intend to expel members for not implementing Responsible Care. Rather, CMA's strategy is to encourage and equip members to implement it. According to Holzman, the CMA follows a three-step process for monitoring progress on Responsible Care and correcting unsatisfactory performance. First, the CMA talks directly to the facility managers (since progress is monitored at the facility level) and gives them deadlines for correcting implementation shortfalls. If this fails, then the CMA contacts senior managers at the firm's headquarters and encourages them to pressure their facility managers to improve performance on Responsible Care. If this does not succeed, then conceivably a firm may lose its CMA membership (*Chemical and Engineering News* 1992).

Lilly: policy dynamics

When the CMA proposed Responsible Care in 1988, Lilly was hesitant in adopting it; or, if forced to adopt, an early internal assessment suggested that it would not implement all of the codes. In particular, some senior managers strongly objected to the code on community awareness and emergency response that required Lilly to develop community outreach programs. They argued that since Lilly was following all the applicable laws, there was little justification for developing such programs. Further, since the public has little technical knowledge of the manufacturing processes, information provided by Lilly could be misinterpreted. Consequently, community outreach programs would create confusion, thereby not serving their intended purpose. Due to the significant hierarchical power of policy skeptics, Lilly did not propose community outreach programs for a few years. Its environmental managers also kept a low profile in discussing this program with employees as well as with external stakeholders.

However, post 1993, Lilly has become a show-piece of successful implementation of Responsible Care. It has also become very active in the CMA on this subject: John Wilkins of Lilly's Corporate Environmental Affairs chairs the CMA's Indiana-Ohio-Kentucky chapter of Responsible Care. Currently, Lilly is implementing Responsible Care codes in the following sites: Carolina (three facilities), Puerto Rico, Clinton (Indiana), Indianapolis (Bulk, Pharmaceutical, and Chemical Process Development, Indiana), Mayaguez (Puerto Rico), and Tippecanoe (Indiana). Even though not required by the CMA, some of Lilly's foreign affiliates are also implementing their respective national versions of Responsible Care. These affiliates included Cosmopolis (Brazil), Kinsale (Ireland), and Speke (UK).Three examples of innovative Responsible Care programs are given below (Lilly 1994c):

(1) Lilly has teamed up with some Indianapolis-based firms for establishing community outreach programs. In addition, many Lilly facilities also undertake solo initiatives. For example, the Tippecanoe facility in Indiana has organized a community forum for discussing its environmental, health and safety performance. This forum meets periodically, typically twice a year. This facility also conducts an annual emergency drill with the county's police, fire, and health services. As a result, the Tippecanoe facility has brought in more transparency in

its decision-making processes and has generated confidence within the local community about its ability to deal with industrial accidents.

(2) Previously, I described how Lilly networked with other Indianapolis-based firms to establish community outreach programs. Lilly, together with these firms, has also established a logistics system for chemical shipments. For example, in the early 1990s, the state of Indiana did not have the necessary statistics for planning shipping operations. These firms, together with their transporters and city/state officials, created a task force for gathering such statistics. This enabled these firms, including Lilly, to schedule shipments in a cost-effective manner.

(3) Lilly has developed a computer-aided logistics system that enables its traffic managers to route shipments through the safest routes. It also closely scrutinizes the safety records and policies of its transport carriers. Further, it has an emergency action plan in case an accident should occur.

As described previously, the CMA recommends that firms annually assess their facility-wise implementation of Responsible Care. As shown in tables 4.9 and 4.10, Lilly evaluates the facility-wise progress on all six codes of Responsible Care. It has invested considerable resources in creating a six-point scale, and auditing facilities on their progress on all six codes. In fact Lilly conducts multiple Responsible Care audits of its facilities every year and reports the annual average scores for every facility in its Environmental Performance Reports. Lilly has set a goal that every facility should score at least 5.0 on each of the six codes. Since progress on Responsible Care is an important input in the evaluation of the facility managers, they have significant incentives to implement this program. In addition, Lilly's Tippecanoe facility is implementing the CMA's pilot program – Management Systems Verification – where external auditors assess a facility's progress on Responsible Care (CMA 1998). This again reinforces the fact that Lilly is currently in the forefront of implementing Responsible Care.

Analysis

Power-based theories are helpful in explaining Lilly's initial reluctance and leadership-based theories in explaining its subsequent enthusiastic implementation of Responsible Care. When the CMA proposed Responsible Care in 1988, some senior managers were hesitant in adopting it due to the code on community awareness and Emergency Response

Table 4.9. *Lilly's internal evaluation of Responsible Care*

	1992	1993	1995*
Community awareness	3.7	3.7	4.5
Emergency response	4.5	4.5	4.7
Pollution prevention	3.6	3.7	4.4
Process safety	3.6	4.0	4.0
Distribution	2.7	3.3	n.a.
Health and safety	4.3	4.7	4.5
Product stewardship	2.0	4.4	n.a.

Notes:
Scale
1 = no action has been taken
2 = evaluating existing experience
3 = developing plans
4 = implementing plans
5 = plans have been implemented and are in place
6 = reassessing the existing plan
* These scores represent the mean of scores of six Responsible Care facilities.
Source: Adapted from Lilly (1992b, 1995c).

that required developing community outreach programs. They argued that communities living in the vicinity of any facility often have little technical knowledge for appreciating how manufacturing processes work. Hence any sharing of technical information may lead such communities to make unwarranted conclusions about the safety aspects and environmental impacts of a facility's operations. As discussed previously, establishing community outreach programs was opposed by many leaders in the chemical industry as well. As Simmons and Wynne observe:

Fundamental to the identity of the chemicals sector is its sense of being a science-based industry. This is deeply ingrained in the industry's culture and belief about the validity and authority of science frame its view of outside groups. These beliefs are reflected in the argument that has been made to legitimate the *industry's claim to self-regulation – that its unmatched knowledge and expertise make the industry's own experts the people best suited to audit and regulate the environmental effects of its activities*. (1993: 218; italics mine)

Thus, even though Lilly had to formally adopt Responsible Care, it kept a low key approach, especially on implementing the code on community awareness and emergency response. In 1993, after about five years of formally adopting Responsible Care, when one of the main opponents (a senior manager) of this program retired, policy supporters spotted a window of opportunity. They were in a better position to convince the

Table 4.10. *Responsible Care: implementation status*

Site	1995 average*	1991 average*	Percent of practices in place in 1995
Carolina	4.1	2.7	28
Clinton	5.0	3.3	73
Indianapolis	4.7	2.7	61
Mayaguez	4.7	2.7	62
Tippecanoe	5.1	3.3	66
Total	4.7	2.9	58

Notes:
Scale
1 = No action has been taken
2 = Evaluating existing experience
3 = Developing plans
4 = Implementing plans
5 = Plans have been implemented and are in place
6 = Reassessing the existing plan
* Since Lilly conducts multiple audits every year, average of all audits scores is being reported.
Source: Wilkins (1996).

remaining skeptics of the benefits of implementing Responsible Care. Particularly, policy supporters highlighted the tremendous benefits, even though non-quantifiable, of establishing community outreach programs in terms of generating goodwill within the CMA and among local communities, regulators, and environmental groups.

Why and how did policy supporters succeed in convincing policy skeptics of the benefits of being in the forefront of implementing Responsible Care? Specifically, what benefits accrue to Lilly for maintaining a high profile on this subject within the CMA? As I have discussed before, the CMA is a significant player in policy discussions on proposed laws and regulations or any issue that impacts the chemical industry. Consider these two examples. The EPA has proposed to increase the number of chemicals to be reported under the Toxic Release Inventory program. This has significant cost implications for the chemical industry since they will come under stakeholder pressure to reduce their emissions of the chemicals added to the TRI list. The CMA has challenged this proposal in a federal court on the grounds that the new additions do not constitute significant threats to public health and therefore should not be clubbed with the original TRI chemicals. It appears that eventually the EPA, the CMA, and other parties will compromise on this subject; the EPA will perhaps scale down the list of additions. Firms having a crucial say in the running of the CMA will have better chances of influencing the eventual outcome in their favor.

The second example pertains to the CMA's response to the public debate generated by the book, *Our Stolen Future* (Colborn 1996). This book suggests that certain chemicals may disrupt the normal functioning of endocrine glands, thereby leading to problems such as sterility. Consequently, many environmental groups have called for a ban on the use of such chemicals. Since evidence on this subject is inconclusive, the CMA has committed $850,000 for further research into the causes of endocrine disruption. Again, the main actors in the CMA will play a crucial role in shaping the CMA's response to this controversy that has bearing on the future of the chemical industry. These examples suggest that firms indeed have significant incentives to gain credibility within the CMA, thereby influencing its agenda. However, since such benefits are not quantifiable, policy supporters had to invest significant efforts to convince the skeptical managers of the potential payoffs of implementing Responsible Care.

One of the concerns of policy skeptics was that implementing Responsible Care would be very expensive. Since it is difficult to assess the adequate level of expenditure on programs such as community outreach, Lilly would have a propensity to overspend. In response to these concerns, Responsible Care supporters developed innovative methods for controlling community outreach program costs. Eli Lilly established a network with other Indianapolis-based firms such as Reilly Industry, National Starch, and Dow Elanco each having facilities in close physical proximity. These firms jointly undertake initiatives such as community outreach activities and conducting science demonstrations at local elementary schools. Thus these firms share costs as well as provide a superior program by pooling their talents (*Chemical and Engineering News* 1993a).

An important task for the policy supporters was to ensure wide-spread internal participation in developing and implementing Responsible Care. Specifically, there was a concern that Responsible Care should not be viewed as a corporate program over which facilities had little control. To meet this objective, a Responsible Care group was established with representatives from various facilities. This group meets once every two months to share experiences in implementing existing Responsible Care initiatives and developing new ones. Establishing this group has resulted in unprecedented enthusiasm among facility-level managers for implementing Responsible Care.

Policy skeptics also argued that there was little clarity about the expectations from facility managers on Responsible Care. This was a crucial issue since most of the Responsible Care programs were to be implemented at the facility level. Policy supporters handled this concern by proposing a six-point scale (tables 4.9 and 4.10) to assess progress on the various codes of Responsible Care. Further, they also established a system of regular internal audits, whereby facilities could have specific and timely feedback on their progress in implementing this program.

To conclude, an important learning from this case is that individuals matter, not only in implementing particular programs, but also in opposing them. Structurally advantaged individuals can ensure the implementation of unpopular policies or thwart the implementation of popular policies. Thus, power-based processes can work both ways: in having policies implemented and in having policies jettisoned. Power-based processes therefore explain Lilly's initial hesitation in implementing all codes of Responsible Care.

Leadership-based processes explain why Lilly successfully implemented all codes of Responsible Care, even though the benefits of such policies were not quantifiable. Once the key opponent of Responsible Care retired from Lilly and was not in a position to influence its environmental policymaking, policy supporters succeeded in generating consensus, especially on implementing community outreach programs. They convinced the skeptics that adopting all codes of Responsible Care would serve Lilly's long-term interests, even though such policies may not generate quantifiable profits. They highlighted that aggressive and visible adoption of Responsible Care would increase Lilly's credibility within the CMA, and with the regulators, environmental groups, and local communities. Credibility within the CMA was particularly important, given the CMA's preeminent position as the representative of the chemical industry in most environmental policy debates. Policy supporters won the trust of skeptics by innovating strategies for controlling costs and enhancing the quality of Responsible Care initiatives. They ensured that no significant group within the company perceives itself as a loser from implementing this program, and hence have incentives to oppose it. This was important as many Responsible Care programs, especially on community outreach, are quite revolutionary in that they expose managers involved in extremely technical tasks to public scrutiny. Thus, a consensus mode of implementation ensures that technical experts do not feel threatened by imposition of such programs, and, at the same time, the firm meets expectations of its external stakeholders in terms of sharing information on its internal functioning.

4 "Green products"

This section focuses exclusively on policy dynamics within Baxter on the marketing of green products. Green products are those that specifically aim to deliver enhanced environmental quality as one of their key product benefits (Coddington 1993). Given the nature of Baxter's product portfolio, it had opportunities to offer such products, and the presence of certain key managers within the firm ensured that such opportunities were tapped. Some products used vast quantities of packaging which imposed significant disposal costs on Baxter's customers.[7] Environmentalists were also concerned about the environmental implications of excessive packaging. Due to the vision of key managers, Baxter aggressively reduced packaging, redesigned its products, and satisfied latent customer need for environmentally friendly low-disposal cost products. These strategies were not justified by using formal investment criteria and yet they did not result in intra-organizational conflicts. Hence, leadership-based explanations best explain Baxter's decision to market green products.

Given Lilly's product portfolio, it does not have a business rationale, in the short or the long run, to manufacture and market green products. The reason is that environmentally friendly features in ethical or prescription drugs (Lilly's main line of business) do not motivate doctors to recommend them to their patients; the cost of treatment, efficacy, and lack of side-effects are the key motivators in brand choice. Thus, developing green ethical drugs or portraying its existing drugs as green provides meager quantifiable or non-quantifiable benefits to Lilly. It is important to reiterate that this book examines cases where firms have adopted policies that are perceived by managers to provide non-quantifiable benefits that accrue in the long-run. Thus, although such policies cannot be defended by employing capital-budgeting, managers believe that they are important and beneficial for the firms' survival and prosperity. This book *does not* make an argument that managers advocate policies that provide little benefits, quantifiable or non-quantifiable, to their firms.

The greening of firms can imply both the greening of their value-addition processes and of their products. An important question therefore is: how does greening influence the purchase decisions of customers? To

[7] Customer means the product's purchaser while consumer means the product's user. For a firm marketing final products, distribution chains are its customers, and individual actors purchasing from these outlets are its consumers. For industrial and intermediary products, where one firm sells to another, the term customer is more applicable. For example, if Baxter sells dialysis machines to health-care centers, these healthcare centers are Baxter's customers.

explore this subject, this book examines two dimensions of greening. First, verifiability of greening by customers or actors external to firms. Second, importance of greenness in influencing purchase decisions. Apropos the first dimension, customers often have little information on this subject, and, consequently, they find it difficult to factor in environmental aspects of firms' processes and products in their purchase decisions. Such information asymmetries therefore lead to market failures.[8] To mitigate market failures, policy interventions take place at various levels. Governments enact laws requiring firms to report on aspects of their environmental program. For example, under the Toxic Release Inventory Program, firms with manufacturing facilities in the US are required to report annual releases of specified toxic chemicals. Likewise, firms seek to mitigate market failures at the industry level by sponsoring industrial codes of environmental self-regulation such as the CMA's Responsible Care program. Firms also seek to mitigate information asymmetries at their level only. For example, they obtain environmental certifications such as ISO 14000 that are awarded by credible external auditors. All such measures, whether at the macro, industry, or firm level, seek to provide credible information on the greenness of firms and their products. Consequently, they help customers to factor in firms' greenness in their purchase decisions.

The second dimension I examine is whether customers consider greenness of firms an important factor in their purchase decisions. Customers base their purchase decisions on a variety of factors such as price, performance, availability, and reliability of products as well as the corporate image of firms that are manufacturing and marketing them. Consequently, the level of greenness of firms and their products is one of the many factors influencing the purchase decision; customers may or may not prioritize it over other factors.

As discussed in table 4.11 below, based on the two dimensions – verifiability of greenness and importance of greenness – we can identify four categories of situations.

Cell 1 represents market situations in which customers not only find it difficult to verify firms' claims on greenness, they also do not attach much importance to them. For example, customers will find it difficult to verify whether a firm has replaced its underground storage tanks with expensive

[8] Based on degrees of information asymmetries, goods can be classified into three categories: search goods, experience goods, and post-experience goods. Consumers can determine the characteristics of a: (1) search good prior to purchasing it; (2) experience good after purchasing it; and (3) post-experience good only with a lag after consuming it. Thus, information asymmetries (and concomitant market failures) are most severe in post-experience goods and least, if at all, in search goods (Weimer and Vining 1992).

Table 4.11. *Impact of greening on customers*

	Customers cannot verify greenness of firms' policies or of their products	Customers can verify greenness of firms' policies or of their products
Environmental issues are unimportant in influencing customers' purchase decisions	Beyond-compliance policies such as removing underground tanks do not influence brand choice of customers of anti-depressants or heart valves (Cell 1)	Adopting ISO 14000 does not influence brand choice of customers of anti-depressants or heart valves that are located in the US. (Cell 4)
Environmental issues are important in influencing customers' purchase decisions	These situations are unlikely to occur. If firms recognize that customers value greenness, then most of them can be expected to provide credible information on it (Cell 2)	ISO 14000 certification is required for selling in some European countries. Many US-based hospitals buying medical products from Baxter value green attributes of such products (Cell 3)

above-the-ground tanks, whether it conducts internal environmental audits, or whether heart valves are bio-degradable. Importantly, even if such claims are verifiable, the adoption or non-adoption of such green policies would not influence some customers' brand choice. Consider an individual wanting to purchase an anti-depressant. It is unlikely that this individual would purchase Prozac, an anti-depressant produced and manufactured by Lilly, simply because Lilly has removed underground storage tanks from its facilities and replaced them with expensive above-the-ground tanks, or because Prozac is bio-degradable, or because it is packaged in recycled material. This individual probably has little access to information on Lilly's environmental policies. And, even if the customer has such access, this knowledge may not significantly affect his/her purchase decision. For an anti-depressant, the customer may focus predominantly on Prozac's performance and cost of treatment, giving little importance to Lilly's environmental policies or to Prozac's environmental impact. Similarly, customers wanting to purchase heart-valves will probably not buy them from Baxter just because they are bio-degradable, or because Baxter invited Arthur D. Little (ADL) to help it define state-of-the-art environmental standards, and then audit its environmental programs against them.

Cell 4 represents market situations in which customers can verify levels of greenness, but they do not consider it important in choosing brands. For example, firms may have the ISO 14000 certification, be members of Responsible Care, or subscribe to the EPA's 33/50 program. All these are examples where the greening of firms can be verified. However, this greening may be unimportant for customers wishing to purchase anti-depressants or heart valves.

If customers do not attach importance to the greenness of firms' policies or products (Cells 1 and 4), supporters of such environmental policies within firms probably do not justify them in terms of their impact on customer goodwill. As discussed in previous sections of this chapter (underground tanks, the 33/30 program, and Responsible Care), the adoption of Type 2 policies is best explained by understanding how policy supporters managed to convince the organization of the long-term benefits of these policies. These benefits include reducing environmental risks, and having the goodwill of regulators, local communities, and industry-level organizations.

Cell 2 represents market situations in which customers consider the greenness of firms or their products important but they have little access to verifiable information on this subject. Consequently, customers have to rely on claims made by firms or on media reports. Such situations are unlikely because if firms recognize that customers value greenness, they can be expected to make efforts to provide information on this subject. If anything, firms are often perceived to be over-stating the achievements of their environmental programs.

Cell 3 represents those situations in which customers are able to verify levels of greenness and they also consider greenness important in influencing their brand choice. For example, purchase decisions of some European customers are critically influenced by whether or not a supplier has the ISO 14000 certification. Or, some customers may only purchase products having a green dot.

Note that greenness can be certified at the firm level (ISO 14000) as well as at the product/brand level (green dot). This distinction becomes important when thinking about business strategy. Do customers buy products based on brand's attributes only or are their purchase decisions also significantly affected by a company's image and philosophy (Charter 1992)? If purchase decisions are significantly affected by a company's image, firms have incentives to invest in firm-level programs and then become certified. As discussed previously, for certain categories of customers, firm-level certification is indeed very significant.

However, if customers buy brands, and not a firm's philosophy, firms have incentives to green their products, to certify, and to communicate it to their customers.[9] Thus, customers can examine the attributes of such products, make inferences about their environmental impacts, and factor them into their decisions on brand choice. This is the focus of this case: how do firms deal with beyond-compliance policies that influence the attributes of their products, given that these attributes are valued by their customers?

Baxter and Lilly: response

Lilly manufactures and markets ethical pharmaceuticals. Some of Lilly's products such as insulin are over-the-counter (OTC). However, since patients often purchase these products on doctors' recommendations, such OTCs are *de facto* ethical drugs. For ethical drugs, the purchaser is not the decision maker regarding brand choice: patients are the purchasers and their doctors are the decision makers. Research suggests that among various attributes of an ethical drug, doctors consider its efficacy most important; after all their professional reputation depends on the drug's performance.[10] One could argue that other things being equal, the "greenness" of the drug may become important. That is, if a doctor is asked to choose between two ethical drugs that are equal in all aspects such as efficacy, cost, and reputation of the manufacturer, then this doctor may prefer the greener drug. However, such other things are seldom equal. Even though generic drugs such as basic antibiotics manufactured by different firms may be equivalent on most major attributes, by employing various tools of marketing, firms often succeed in differentiating their brand from the rest.

Further, since the health-care industry is highly research intensive, new drugs claiming better efficacy are being continually introduced. As a result, it is difficult to imagine that doctors will face similar situations on brand choice for ethical drugs in any major disease category. Of course, one can argue that doctors may be willing to prescribe greener drugs that

[9] Companies such as Procter and Gamble, General Mills, and Colgate focus on communicating the benefits of their brands. I am not arguing that such brand-focused firms ignore their corporate image. They do not. However, such firms focus most of their advertizing on talking about their specific products. For example, most of Procter and Gamble's advertizing focuses on the superior performance of Tide detergent, the freshness of Ivory soap, or the beauty-enhancing effect of Oil of Ulay. Most consumers probably do not link all these brands to Procter and Gamble. Of course, some other firms focus on corporate advertizing or have a generic brand name for their various products. In particular, Japanese companies have generic brand names. Thus, we have a SONY television, SONY VCR, and SONY Walkman.

[10] I have similar findings from my previous career at Procter and Gamble as a brand manager of the Vicks range of OTC drugs.

are only marginally less efficacious (or marginally more expensive) than non-green drugs, as long as greener drugs are efficacious enough (or cheap enough) for treating patients. Such possibilities certainly exist for only basic drugs such as basic antibiotics or sulpha drugs, and not in Lilly's product categories.

I am not arguing that firms' environmental policies do not impact doctors' brand choice for ethical drugs. They certainly do. Suppose there are extensive media reports on pollution caused by a pharmaceutical firm or adverse environmental impacts from its products. Surely doctors may begin to have negative opinions about this firm and its brands. However, adoption of beyond-compliance policies or greening products will probably not persuade doctors to prescribe the firm's brands. To employ Herzberg's (1966) terminology, for doctors, a lack of negative publicity about the environmental policies of a firm, or the environmental impact of an ethical drug is a "hygiene" factor.[11] However, the presence of environmentally friendly features in ethical drugs or the adoption of beyond-compliance policies by a firm is often not a "motivator" for recommending the brands of ethical drugs manufactured by this firm. Factors such as the cost of treatment, efficacy, and lack of side effects are the motivators' for doctors in brand choice. On this count, Lilly has little reason to market green ethical drugs or to portray its existing drugs as green.

In contrast to Lilly, Baxter markets green products: medical supplies that are purchased by hospitals and other health-care suppliers. Some of Baxter's green products are briefly described below (Baxter 1994d; 1996b).[12]

(1) Access: Baxter and Waste Management, Inc. have formed a strategic alliance to offer Baxter's corporate customers solutions to all their environmental issues. Waste Management, Inc. is the world's largest waste disposal company offering consulting services on waste reduction, recycling, treatment and disposal of waste, etc.

(2) ValueLink: Baxter repackages bulk products to customer

[11] Extending Maslow's (1943) theory, Herzberg develops a theory of work motivation. His study focuses on two work-related factors: those that turned people on and those that turned them off. He labels the former 'motivators'; these factors pertain to the job content such as levels of creativity and recognition. He labels the latter 'hygiene' factors; their presence only prevents dissatisfaction among employees. These factors are related to the job context such as company policies, working conditions, and salary.

[12] Baxter sought to address customers' environmental concerns in other ways as well. In 1995, it established a Customer Waste Council (CWC), a multi-divisional and cross-functional initiative to better respond to customers' environmental concerns. As of 1997, the CWC consists of five teams whose objectives include evaluating the latest technologies in reusable shipping containers; investigating what environmental information consumers are requesting, why they are making these requests, and how Baxter can fulfil them (Baxter 1996b).

specifications, delivers them daily to the patient floors using reusable totes, and recycles their bulk-packaging. Consequently, this eliminates bulk-packaging from customers' waste streams.

(3) Envision Recycling: Baxter provides training to hospital staff to sort plastic intravenous bags, pour bottles, and over pouches. Baxter then collects, transports, and recycles them.

(4) Custom Sterile Packs: Baxter offers customers individualized packs of medical items thereby reducing packaging and saving them disposal costs.

(5) Interwoven: Baxter offers to collect, clean, and deliver disposables such as gowns, drapes and towels so that they can be reused by customers.

(6) Shared-Risk Agreement: The objective is to reduce the amount of products used by Baxter's customers. For every customer, Baxter analyses past patterns of product usage. Then both parties establish a target usage for the future. If the customer uses more than this target, then both Baxter and the customer share the additional costs. If the usage is below target, then both Baxter and the customer share the resultant savings.

Baxter's intra-firm dynamics

Why and how did Baxter decide to market green products? New product launches are typically subjected to rigorous capital budgeting but Baxter did not employ any established procedures to evaluate marketing green products. Power-based explanation is inadequate because marketing of green products has been widely supported within Baxter. There was indeed some initial opposition but policy skeptics did not dig in their heels because they did not view this policy as imposing significant costs on them. Leadership-based theories are most appropriate to explain Baxter's response to marketing of green products. Policy supporters succeeded in convincing their skeptical colleagues that the demand for Baxter's green products is not a passing fad; it reflects fundamental changes in the ways customers prioritize the desired attributes of any product. They pointed out that new laws and regulations (see below) would create incentives for customers to demand products that have less packaging or that do not require polyvinyl chlorides (PVCs) for their packaging. Thus, policy supporters created a consensus about marketing green products, even though they did not employ well-established investment appraisal procedures.

To understand the policy dynamics within Baxter, it is instructive to examine its external environment as well. Many recent environmental laws and regulations have sought to promote the reduction, reusage, and recycling of resources. The objectives of such laws are twofold: to reduce the volume of wastes, particularly solid wastes, that are overwhelming landfills, and to conserve raw materials. One of the strategies has been to enact "take-back" legislation that requires manufacturers or retailers to take-back packaging at no additional cost to consumers. Such laws are intended to create incentives for manufactures to reduce packaging. Europe, in particular, has pioneered such "take-back" legislations. In 1981, Denmark enacted a law requiring manufactures to use reusable beer and soft drink containers and requiring beverage retailers to collect all such containers regardless of whether or not they sold them.[13] In 1991, Germany enacted its own version of "take-back" legislation requiring firms, or their designated agents, to collect and to recycle all packaging. Consequent to a rather successful German program, in 1994 the European Commission issued its own version of "take-back" and recycling directive. In the US as well, pressure has been building for firms to reduce their packaging. As discussed in Chapter 3, in 1991, the Coalition of North Eastern Governors (CONEG) challenged the top two hundred users and producers of packaging in the US to eliminate or to reduce the amount of packaging generated by their companies.

Most of Baxter's green products emphasize packaging reduction which will reduce pressure on their customers' waste streams. This strategy makes business sense because many of Baxter's products require substantial packaging. Note that, in contrast, Lilly markets products that require little packaging. As a result, packaging reduction is almost irrelevant for Eli Lilly.

In response to the CONEG challenge, Baxter committed to a 15 percent reduction in per-unit packaging weight by 1996 with 1990 as the baseline (Baxter 1991b). Baxter's packaging task-force set guidelines on issues such as reducing the use of inks containing heavy metals, modifying Corporate Identity Guidelines for the company logo to allow for the use of recycled paper for office stationary and packaging materials, applying Society of Plastics Industry and American Paper Institute recycling symbols for appropriate packaging, promoting the sale of single package multi-product medical kits, advocating the use of reusable shipping containers, and minimizing the use of chlorine-bleached papers and paperboard in packaging. Thus, policy supporters argued that such initiatives

[13] The Danish legislation was criticized however as a non-tariff barrier. Though the "take-back" clause of this regulation still holds, the European Court of Justice struck down the reusable provision (Vogel 1995).

provide Baxter an opportunity to leverage its packaging reduction expertise for marketing green products.

Policy supporters also highlighted the EPA's (then) recent proposal on stricter air-pollution control standards that would have required an upgrade in existing medical waste incinerators. Hospitals (Baxter's customers) either have their own incinerators or they send their medical waste to a third party for incineration. It was estimated that such an upgrade might cost up to a million dollars per incinerator. Consequently, if implemented, this policy would shut down about 80 percent of medical waste incinerators. Policy supporters argued that this proposal would create incentives for hospitals to demand that Baxter reduce the quantity as well as certain types of packaging (Baxter 1995b), and such demands would be in tune with the general concern among health-care providers about managing medical wastes. As Marshall Abbey, former Senior Vice-President, General Counsel, and Chairman of Baxter's Environmental Review Board, noted:

> Remember, a large share of Baxter's revenue comes from the sale of disposable medical products. As soon as these products are used, they become waste. And . . . waste is becoming a big problem for our customers. So while we are selling products to our customers to satisfy their medical services needs, we are also giving them waste problems. If we are to keep our customers and attract new ones, it is critical that we work to minimize these problems. The future of the company depends on this. (Baxter, 1991a: 3)

In similar vein, while addressing the conference on "Our Environment: A Healthcare Commitment," held in Arlington, Virginia, March 10, 1992, Vernon Loucks, Baxter's CEO noted that:

> [O]ne of the hospital's biggest opportunity involves managing waste. A 500-bed hospital in the US generates abut 1,500 tons of waste each year – three tons per bed – and it can spend up to $350,000 getting rid of it. A teaching hospital of that size typically generates much more waste, and its disposal costs can be well over $1,000,000. But a hospital can whittle down these disposal costs significantly through aggressive programs to reduce, separate, and recycle waste.
>
> Baxter's ACCESS program has shown that this approach really works . . . the 288-bed Children's Memorial Medical Center in Chicago, saved $500,000 in waste disposal costs over three years. Another customer, Thomas Jefferson University Hospital in Philadelphia, established a recycling program that produced a $100,000 savings the first year. (Baxter, 1992b: 4)

Another focus of Baxter's efforts to green its products was reducing the use of polyvinyl chlorides (PVCs) for packaging. Like many other firms in the health-care industry, Baxter uses PVCs to package many of its products. PVCs are plastics derived from mineral oil, natural gas, and rock salt. It is suggested that when incinerated the chlorine in PVCs may produce dioxins. However, there is no statistically proven relationship between the amount of PVCs incinerated and the levels of dioxin produced. Dioxins, the generic name for a group of seventy-five chemicals known as chlorinated dibenzo-p-dioxins, are shown to cause cancer in some laboratory animals. As a result, there is immense pressure from environmental groups for regulating, if not banning, the use of PVCs. The medical waste incinerators are considered one of the biggest sources of dioxins, primarily because they incinerate PVCs. The controversy over PVCs created incentives for these customers to demand that Baxter should reduce, if not eliminate, the use of PVCs to package its products (Baxter 1996a).

A similar controversy has been plaguing di-2-ethylhexyl phthalate (DEHP), a plasticizer for PVCs. Some scientists claim that substances such as DEHP that exhibit estrogen mimicry may substantially reduce sperm counts and cause other reproductive problems. A widely discussed book, *Our Stolen Future* (Colborn 1996), argues along similar lines. Again, though there is no conclusive evidence on the impact of DEHP on reproductive issues, nevertheless, there is intense pressure by environmental groups to regulate, if not eliminate, their use. Some European countries such as Germany and Denmark have already taken a position against DEHP. This controversy again created incentives for Baxter to reduce its packaging and the use of PVCs (Baxter 1996c).

Though policy supporters highlighted that new laws and regulations create incentives for marketing green products, they also invoked normative factors and pressures from non-governmental actors to generate support for these policies. They emphasized that as a health-care firm Baxter was responsible for safeguarding the health of the environment as well. Baxter's CEO Vernon Loucks, almost acting as their chief spokesman, noted:

> So there are good business reasons to invest in the environment. There are ethical reasons, as well. Environmental protection lies at the very core of our social duty. Our industry is dedicated to preserving and improving the health of mankind. . . . Indeed, our industry has a special connection to life that no other industry shares. But we cannot preserve that trust and uphold that bond if we ignore the environmental consequences of our action. *We cannot fight for life at one moment and destroy it the next. We cannot help those at our front door and harm them by the wastes we send out the back.* (Baxter 1992a: 5–6; italics mine)

This book has also suggested an important role for non-governmental external stakeholders in encouraging Baxter to market green products. One way in which such actors exercised influence was by communicating their demands and expectations to Baxter. Actors who supported marketing green products publicized such communications within Baxter, particularly by getting them published in the in-house environmental newsletter *Baxter Environmental Review*. Let me quote from letters written by three different categories of actors: a household using Baxter's product, a hospital, and an investment firm.

Jim and Sarah Tennessen of Menomonie, Wisconsin, sent this letter to Caremark, a division of Baxter:

> Dear Caremark:
>
> I am writing you with a challenge. Caremark has a reputation for innovation and quality. I would like to see Caremark apply this innovation towards preserving the quality of the planet we live on. I know of no home care provider who has any type of recycling effort in place at this time. I see this as a tremendous opportunity for Caremark to take on an industry role, not only improving the quality of life for its customers, but maintaining the quality of life for all mankind. Caremark, with its home delivery network, is well-positioned to capitalize on this opportunity.
>
> In a single week, our four year old's HPN [TPN] therapy creates approximately six cubic feet of waste, consisting of seven 2,000 ml I.V. bags, three 250 ml lipid bottles, ten line sheets, seven vitamin and sterile water vials, 21 syringes, 24 needles, 16 face masks, 67 alcohol swabs, four split bandages, two gauze bandages, one role of tape, plus all the miscellaneous packaging for these items .... (Baxter 1991a: 1)

Similarly, Mr. David Baker, Director of Materials Management, St. John Mercy Medical Center, St. Louis, wrote:

> One of our employees recently asked our president whether he knew the environmental records of the many vendors we deal with. To that end, we would like to request information from your firm about the composition of the products you sell, the functionality of your products and services – anything you do environmentally sound. We would appreciate sharing this information with our employees. (Baxter 1991a: 2)

Scott Fenn, Director of the Investor Responsibility Research Center, a non-profit firm representing 400 investment institutions, including many large investment funds, sent a survey to Baxter seeking information:

> Has your company introduced or modified any product or service in the last three years specifically to address environmental concerns? Please indicate the nature of the environmental or financial benefits the company hopes to achieve (Baxter 1992a: 2).

To conclude, Baxter is marketing green products because it could build on its competencies in recycling and packaging reduction, its customers value the benefits of green products for environmental or profit reasons, and customers believe that Baxter's products deliver these benefits. In contrast, Lilly has few business reasons to market green products since most of its products are ethical drugs whose purchases depend on factors such as efficacy and price. These drugs also have small dimensions and hence there is not much scope for reducing their packaging. In addition, these drugs are single-use products that cannot be repaired, reconditioned, remanufactured, or reused.

Since Baxter has not done any investment analysis to justify marketing green products, efficiency-based theories offer little help in understanding Baxter's internal policymaking on this subject. Baxter markets a range of green products and this book has explained the adoption of this policy by employing leadership-based explanations. To persuade the organization to market green products, policy supporters highlighted various benefits of this policy.

First, they emphasized its normative aspects: products of a health-care company should not damage the health of the environment. Second, since many non-governmental actors were asking for information on Baxter's green products and communicating their expectations on this subject, the policy supporters highlighted the goodwill benefits of marketing green products. Third, and most important, policy supporters emphasized that marketing green products will eventually be profitable for Baxter since such products provide quantifiable savings to customers. As discussed earlier, by emphasizing recycling and packaging reduction as key benefits, Baxter's green products help customers in lowering their waste disposal costs. Further, these products also proactively reduce PVC incineration.

These benefits are valued by Baxter's customers in light of the EPA's proposal on upgrading air-control standards for medical waste incinerators. If implemented, this policy would require expensive refitting of such incinerators. Consequently, many of them will shut down and the remaining ones will begin charging much higher prices for incineration. Therefore, the policy supporters emphasized that the demand for green products will be primarily driven by an unmet need: customers' desire to reduce medical waste disposal costs. However, these benefits were not obvious to many managers within the organization. Thus, adoption of policies on marketing green products is best explained by focusing on the role of the policy supporters who explained the business potential of such products, and as a result, generated consensus on adopting such policies.

5 Environmental audits

Both Baxter and Lilly regularly audit their environmental programs. They have internal audits teams that audit these programs at various levels: corporate, division, and facility. Baxter also invites external actors to conduct some of its environmental audits. It invited Arthur D. Little (ADL) to help define state-of-the-art environmental standards and then audit its environmental program against these standards. Further, Baxter even makes such audit findings public.

Environmental audits have become controversial in recent environmental policy discussions. Enforcing environmental laws is a key challenge for state and local government agencies. Often, these agencies are inadequately staffed. Further, employee motivation is low since they are not well-paid and they receive little training for upgrading their skills. In addition, the sheer complexity of most environmental laws and regulations requires that these agencies devote significant proportions of their resources to processing permit applications from the regulatees rather than monitoring and enforcing compliance. Consequently, even in instances of alleged non-compliance, state and local-level regulators are often handicapped by lack of resources to document their cases against the alleged violators.

Environmental regulators have two routes by which to address this issue. First, to relax the resource constraint, they could ask for increases in their agencies' budgets. This would enable them to hire more people, pay them higher wages, and invest in their training. As a result, agencies would potentially become more effective in enforcing environmental laws. This route operates from the supply side: that is, increasing the supply of regulatory services by budgetary increases. This route, though attractive, is difficult in an era of downsized governments, increasing unfunded mandates to the states, and constant criticism of the adversarial relationship between regulators and industry (Chandler 1980; Marcus 1984; Vogel 1986).

The second route to monitor and prosecute non-compliance operates from the demand side: reducing the demand or decelerating the increases in demand for enforcement and monitoring services in the face of declining, constant, or gradually increasing supply. Such demands are being articulated by a variety of actors such as common citizens, environmental groups, media, and legislators. To meet the demand for these services, regulators can propose to create new institutions (such as self-audits) that generate incentives for the regulatees to self-monitor and voluntarily comply with the laws.

The idea behind environmental audits is that if regulatees can create credible management systems to monitor their own compliance and to correct the management systems that led to non-compliance, the demand for enforcement from the regulators will be reduced. After all, a cleaner environment is the end product of enacting complex and stringent environmental laws and setting up regulating agencies. If regulatees can achieve this end product by themselves, then there would be less demand for external enforcers. As a secondary benefit, since many regulatees feel threatened by the idea of regulators knocking at their doors and peering at their compliance records, this plan will also lessen the confrontational relations between industry and government. Thus, this self-monitoring route seems to be win–win for the regulators and the regulatees.

Unfortunately, the situation is complicated by many issues. Suppose a firm conducting a self-audit of its environmental program discovers violations of environmental laws. Should this firm report these violations to regulators? Will these audit findings be self-incriminatory: will the regulators use them as evidence against this firm in civil, criminal, or administrative proceedings? If so, firms have few incentives to conduct self-audits and to report violations to regulators.

Further, suppose this firm does detect violations and does not report them to regulators or make this information available to the public. Can such information be subpoenaed? Again, firms have few incentives to undertake internal audits and to create evidence that could potentially be used against them.

In the early 1990s, some state legislatures led by Oregon, began taking a serious look at passing laws that would create incentives for firms to undertake environmental audits.[14] By the middle of 1995, about twenty states had enacted legislation that encouraged firms to undertake environmental audits. The passing of such legislation was not well-received by the EPA, the Department of Justice, most environmental groups, and some sections of the legal community. The EPA threatened to withdraw the enforcement authority of the delegated federal laws, particularly the Clean Air Act and the Clean Water Act, from states that had passed such legislation (*New York Times* 1996a) and interestingly, some regulatees like Intel Corporation supported the EPA's stand. Needless to say, the EPA was widely criticized. Senator Enzi chided the EPA by noting:

It is particularly galling to hear the EPA proclaim these state environmental laws ineffective. Given the federal government's behavior, how can they possibly be a brilliant success? Determined to extinguish these laws, the EPA and the Department of Justice have stated their intention to proceed with investigations of companies that use environmental self-audits – rendering these state laws moot. The feds have also threatened the states with loss of what little control they have in

[14] Most laws stipulated that the attorney–client privilege is inapplicable if: (1) the court determines compelling reasons to disclose such information; (2) the firm has abused this privilege for a fraudulent propose; or (3) this information suggests immediate danger to public health or the environment outside the facility.

the administration of the Clean Air Act and Clean Water Act costing them millions of dollars, if they don't stop their environmental audits. This is blackmail. . . . *And it is effective blackmail indeed, for until these audits laws are allowed to function as intended, business will never use them to their full promise.* (1997: A 18; italics mine)

The EPA's critics also challenged its opposition to granting attorney–client privilege for self-audits by regulatees. They pointed out that the Federal government grants such privilege for self-audits in areas such as the Equal Credit Opportunity Act and the Federal Aviation Administration. Responding to such criticism, in December 1995, the EPA issued guidelines on self-policing by regulatees. Though it supported self-audits, the EPA reiterated that it would not grant attorney–client privilege to regulatees. However, it promised not to file for punitive damage (gravity component) if the regulatee notifies the appropriate governmental agency within ten days of discovery of the violation, undertakes follow-up action within sixty days of the audit, provided that there were not any repeat violations at the facility during the following three years nor patterns of violation by the company during the following five years. However, the EPA might still pursue the "economic gain component," that is, recover the gains accruing to the regulatee by violating laws. The EPA also said that it would not pursue criminal action if the violation posed no serious harm, if the management did not conceal or condone such violations, and if such violations were not a result of deliberate managerial blindness (Baxter 1996a; *New York Times* 1996a).

As suggested earlier, most environmental groups opposed granting attorney–client privilege for environmental audits. They argued that such privilege has the potential of being misused by firms: firms can escape prosecution simply by acknowledging that they had violated environmental laws. Consequently, firms would have few incentives to invest resources in implementing laws. Raena Honan, Legislative Director of the Grand Canyon Chapter of the Sierra Club described self-audits as the "Bhopal Bill." She argued that:

Any self-reported violation would be secret and immune from civil plus criminal penalties for anything less than a second-degree murder. . . . We are here at the local level watching out for the public's interest because states do not have a record of protecting their citizens. They have a record of granting corporate welfare and pandering to special interests. (*New York Times* 1996b: A26)

In his letter to the EPA's Administrator Carol Browner, Carl Pope, Executive Director of the Sierra Club, suggested that his group may sue the EPA if the agency did not force the states to comply with the federal environmental laws (*New York Times* 1996a). Some sections of the legal community were also very critical of granting attorney–client privilege to regulatees. Edward Heilig, an Assistant District Attorney in the environmental crimes unit in Suffolk county, New York, noted that:

> All they have to do is to stamp their documents with the words "audit privilege," and we won't be able to look at them (*New York Times* 1996a: A16a).

As a result, Heilig argued that his office would be denied access to information required to investigate environmental crimes. The National Association of Attorneys General identified some legal pitfalls as well (Morandi and Pascal 1995). They criticized the fact that the onus of proof had been shifted to the regulators who were now required to demonstrate that a violation had occurred before receiving access to audit reports. Further, the judges and not the juries (juries tend to be less sympathetic towards firms) would determine whether a firm abused its audit privilege. As a result of these hurdles, regulators would have to devote more resources towards legal battles for having access to the necessary documents on environmental violations. Consequently, the major objective of the new audit laws – reducing enforcement responsibilities of the regulators – would not be served.

To summarize, though some states have enacted legislation granting attorney–client privilege to environmental audits, the EPA, many environmental groups, and some sections of the legal community vociferously oppose such provisions. The EPA does not recognize state laws that grant such privileges. As a result, even if states grant these privileges, the EPA can still have access to the audit information and employ it as evidence against the regulatees for violations under federal laws. Thus, firms have few incentives to establish environmental auditing systems, to undertake audits, and to report violations that have been detected by such audits. Further, if they conduct audits, we expect them to focus on management system audits and not on compliance audits (discussed below); that is, they will focus on assessing whether appropriate environmental management systems are in place, and not on whether any law has been violated. Even if they document that some management systems are not in place, this will not necessarily create self-incriminatory evidence.

Baxter and Lilly: policy dynamics

Baxter and Lilly are committed to conducting environmental audits and have established active programs to audit their facilities, divisions, and corporate groups. Eli Lilly's *Environmental Policy and Guidelines* considers auditing an integral part of Lilly's environmental policy:

Table 4.12. *Internal audits performed at Lilly*

	Lilly facilities	Non-Lilly facilities
1992	8	15
1993	8	23
1994	12	12
1995	15	5

Source: Lilly (1997b).

The company will regularly assess and report to management and the Board of Directors on the status of its compliance with this policy and with environmental laws and regulations. (Lilly 1996b).

Likewise, Baxter's *Environmental Policy* states that:

Corporate environmental personnel, division, and facilities will provide coordinated, effective environmental training, awareness and audit programs wherever relevant (Baxter 1993/94).

Both firms conduct two types of audits: compliance and management systems. Let me give an example to differentiate these audits. Assume that, while auditing a facility, the auditors discover an unmarked drum containing hazardous wastes. If this were a compliance audit, the auditors would inquire whether there are any law(s) that require drums containing hazardous substances to be appropriately marked. If so, which laws, and who is to be held responsible for this violation? In contrast, if this were a management systems audit, the auditors would inquire whether the facility has management systems to ensure proper labeling of drums containing hazardous wastes. If so, why did they fail, and how can the system design be modified to prevent reoccurrence of such failure?

Lilly's environmental auditing program was formally initiated in 1989. In that year, Ralph Hall of the Corporate Environmental Legal group led Lilly's first internal audit that focused on the compliance of the Tippecanoe facility with the requirements of the Resource Conservation and Recovery Act (RCRA). During 1989–1994, Lilly focused primarily on compliance audits. Since 1994, systems audits have also gained prominence. As shown in table 4.12, since 1992, Lilly has also been auditing its suppliers, waste-treatment sites, and third-party manufacturers. In 1996, Lilly also began auditing its non-US facilities. As of 1997, Lilly relies on internal audits only. Lilly's audit teams are comprised of managers from its corporate environmental groups (Environmental Affairs and Environmental Legal) and environmental managers from plant sites outside of the sites being audited.

As discussed in chapter 3, in the late 1980s, Baxter began revamping its environmental program. Baxter initiated its environmental audit program in 1988. It hired Arthur D. Little (ADL) to assist the Environmental Legal Affairs group in developing an audit protocol. ADL is well known for its expertise in environmental auditing; it has been doing such audits since 1977 and conducts about 400 environmental audits annually. Once the protocol was developed, Baxter tested it by conducting pilot audits in three facilities: Bentley facility in Irvine, California, Pharmaseal facility in Jacksonville, Texas, and the Dade-East plant in Miami, Florida. Overall, the protocol was found to be excellent, requiring only minor changes (Baxter 1988). As a result, the environmental protocol was adopted firm-wide.

In 1990, Baxter took another critical step towards strengthening its environmental auditing program. As discussed in chapter 3, Baxter's senior management decided that Baxter would have a state-of-the-art environmental program. It invited ADL to help them define state-of-the-art environmental standards for firms in similar risk categories, and then to audit Baxter's programs against these standards. If a facility met these standards, it was certified as having achieved state-of-the-art standards. ADL also noted that since state-of-the-art standards are dynamic, periodic assessments are required to recalibrate them, and then to evaluate whether facilities meet the new standards. Consequently, ADL audited again in 1993 and 1995. In 1995, it co-audited with Skadden Arps.

Baxter also reported a summary of audit findings in its annual *Environmental Performance Report.* For example, it reported that in 1995 there were 464 audit findings. Of these, fourty seven were considered major items defined as serious deficiencies that pose significant risks to the company: failure to comply with an important regulatory requirement, or lack of a key environmental management system (Baxter 1996a: 18).

Analysis

Power-based and leadership-based theories together explain: (1) why Baxter and Lilly undertake environmental audits, and (2) why Baxter does, and Lilly does not, invite external auditors to undertake such audits. Baxter's adoption of environmental audits, especially its invitation to ADL to help define state-of-the-art standards, and the accelerated schedule of its implementation, was met with internal opposition. The Divisional and facility managers were uncomfortable in committing to ambitious targets, and to inviting external auditors. Given this persistent opposition, and at the same time, a commitment to implement these policies in a short time schedule, policy supporters resorted to having these policies imposed on the organization. Baxter's response to environmental audits is therefore best explained by power-based explanations.

Lilly's adoption of environmental auditing policies is best explained by leadership-based theories. Policy supporters managed to create a broad degree of organizational consensus on a new way of evaluating performance on environmental issues. Their strategy was to convince skeptics, whether at facilities or in the corporate groups, of the long-term benefits of having strong internal environmental audit programs, notwithstanding the EPA's opposition. Importantly, skeptics did not view environmental audits as inflicting losses on them. Consequently, they did not have incentives to dig in, and resolutely oppose these policies.

Lilly's environmental audit program has survived and prospered for many reasons. First, this program has had very committed supporters, particularly Joan Heinz of the Corporate Environmental Legal group, and John Wilkins and Jonathan Babcock, both of the Corporate Environmental Affairs group. These individuals ensured that environmental audits are employed not merely as tools of corporate oversight, but also as tools of system improvement. As one of the managers observed, the audit group is careful not to appear to the facilities as if they are the internal EPA. As a result, facility managers have not felt threatened by the prospect of having their facility-level environmental programs audited.

Second, these managers convinced the organization that Lilly's environmental management systems were fairly robust and that they do not expect audits to uncover any major violation that the EPA or other regulators would not have discovered in the normal course of investigation. As discussed earlier, since the EPA does not recognize attorney–client privilege for environmental audits, firms are wary of conducting audits and creating self-incriminatory evidence.

Third, these managers emphasized that, even though audits were opposed by the EPA and environmental groups, in the long run these actors would begin to appreciate their usefulness. Thus, these individuals succeeded in convincing skeptics within Lilly that a proactive policy to establish an internal auditing program may have some short-term risks but significant long-term payoffs.

The critical watershed was the first audit of the Tippecanoe facility conducted in 1989. Since it revealed certain discrepancies and shortcomings (primarily caused by inadequate staffing), it provided impetus to supporters of environmental programs within Lilly to ask the Environmental Management Committee (EMC) for more resources to strengthen audit programs. Robert Williams, the Vice-President of

Environmental Health and Safety (EHS) and Quality Control/Quality Assurance (QC/QA), and Daniel Carmichael, Deputy General Counsel, were key actors in convincing the EMC to institutionalize environmental auditing. In the highest positivist tradition, they argued that Lilly should be able to measure its compliance with its *Environmental Policy and Guidelines* and that environmental audits were useful tools for this task. Thus, policy supporters were able to leverage the outcome of the first audit and convince the policy skeptics of the benefits of institutionalizing this program.

Fourth, the environmental audit group has constantly explored ways to make environmental audits less burdensome for Lilly's facilities. Environmental auditing is a cumbersome and documentation-intensive process primarily because of the complexity of environmental laws. As a result, facility managers spend a significant amount of their time preparing for audits and implementing the findings of the auditors. Facilities are subjected to other audits as well, such as financial, QC/QA, and health and safety. Consequently, one of the major concerns of policy skeptics was that environmental audits would create unnecessary diversions for the facility managers, thereby impeding them from focusing on meeting production targets. In response to these concerns, the environmental audit group has been very sensitive to the needs of the facilities when scheduling their audits. To minimize disruption in the facility's routine operations, they communicate their audit check lists in advance. Currently, they are exploring with the QC/QA group whether joint environmental/QC/QA audits can be conducted. All such actions signal to the organization the sincerity and constructive intent of environmental auditors. As a result, the policy supporters generated a broad degree of organizational consensus on the usefulness of environmental audits, notwithstanding the hostile stance of the EPA towards it.

Unlike Lilly, why did Baxter's policy supporters rely on power-based processes? Inviting external auditors was clearly a very controversial issue within Baxter. Policy supporters argued that inviting ADL was a demonstration of Baxter's commitment to making its environmental record accessible to external actors and as a testimony of confidence in its environmental programs. They further suggested that since ADL is the oldest, and probably the most established environmental consulting firm, its input in defining state-of-the-art standards will be most helpful. Consequently, external actors will view the state-of-the-art standards defined by ADL as credible and not self-serving. ADL's audit will also provide a useful external perspective on how Baxter could improve its environmental performance.

However, the proposal to invite ADL was internally opposed on four counts. First, the skeptics were concerned about the fuzziness of attorney–client privilege for environmental audits. Second, some managers were uncomfortable with the notion of state-of-the-art standards and committing their facilities to these standards on an accelerated implementation schedule. Some of them questioned the very possibility of defining state-of-the-art standards. Third, some facility managers viewed internal audits as yet another coercive tool employed by the corporate office to make them fall in line. Like many other large firms, Baxter has a tradition of contestation between the corporate office and the facilities. Environmental audits were yet another arena of such contestations. Fourth, since the cost of hiring external auditors was to be eventually borne by facilities, facility managers questioned the need for inviting external auditors at all.

The policy supporters realized that it would be difficult to generate quick consensus within Baxter on environmental audits. I believe a key reason was that the pace and extent of change in environmental programs required to implement state-of-the-art standards were significant. In the mid 1980s, Baxter had a faltering environmental program, and in the 1990s, it decided to have the best program in the industry. Instead of slow and consensual change, policy supporters wanted rapid change. Since such change was resisted, it had to be mandated from the top management. In October 1990, Vernon Loucks, Baxter's CEO, in his address to Baxter's annual internal environmental conference, observed that:

> Our overall goal as stated in this policy is this: to develop an environmental program that will be considered state-of-the-art among the Fortune 500 companies. By state-of-the-art we mean among the best in terms of day-today problem solving and compliance, and in managing our short- and long-term environmental risks. We also mean among the best at minimizing the generation and discharge of waste and other adverse impacts on the environment. Our focus is not just on our operations, but on our products and services as well. (Baxter 1990b: 3)

Loucks went on to emphasize that dissent would not be tolerated. He said:

> *If anyone doubts we are serious about our new policy and goals, let me put that to rest now.* We will monitor the company's environmental program, and the program of each division thorough and annual state-of-the-program report. I have requested that this report be presented to the Senior Management Committee and our board of directors. *Let me assure you, inadequate progress will not be tolerated.* (Baxter 1990b 3–4; italics mine).

To reinforce Vernon Louck's message, during the 1991 annual environmental conference, Marshall Abbey, the Senior President, General Counsel, and Chairperson of the Environmental Review Board observed:

Why go to all this trouble when we could probably get by with much less? The answer to this question lies in another question: what kind of company do we want to work for?

Do we want a company that is concerned with people's health and their right to live in a healthy environment? Do we want a company that stands apart from polluters and violators of environmental law; a company that is doing something about the environment, not just giving a lip service?

Vern Loucks wants this kind of company. The Environmental Review Board does too. And if that's the kind of company you want, I think you do, then you must find one like Baxter that is pressing for state-of-the-art in the environmental area. That's what our goal is all about. *And that's my answer to the skeptics.* (Baxter 1992b: 4; italics mine).

Highly visible support from the CEO and the Chairman of the ERB clearly signaled that Baxter's senior management attached high priority to the environmental audit program. As a result, Type 2 policies on environmental audits, both internal and external, were adopted inspite of opposition from policy skeptics. Again, the learning from this case is that if policy supporters are convinced of the usefulness of a given program, and have the organizational clout to impose their preferences, the program would be implemented.

To conclude, although there was legal controversy on the status of attorney–client privileges for environmental audits, Lilly and Baxter established strong audit programs. In Lilly, policy supporters succeeded in inducing consensus because the pace of change was gradual and audit standards were defined internally. Importantly, their firm was committed to implementing all applicable environmental laws, had established systems, and had invested resources for achieving them. Consequently, skeptics bought into the argument that environmental audits were not expected to uncover major violations that could create problems for them.

In Baxter, the environmental audit program was adopted through a power-based route. Some managers were skeptical about the pace of implementing the state-of-the-art standards, and that these standards were being defined with the help of ADL. Since policy supporters did not succeed in generating a quick consensus on these issues, they resorted to having the policy imposed by the top management.

6 ISO 14000 environmental management standards[15]

So far the book has examined cases where Baxter and Lilly adopted Type 2 policies due to leadership-based and/or power-based processes.

[15] This section draws on Prakash (1999a).

This section examines a policy where neither Baxter nor Lilly adopted a Type 2 project – the International Organization for Standardization's (ISO) ISO 14000 environmental management standards. For the time period of this book's study (mid 1975–1996), both Baxter and Lilly adopted a wait-and-see policy towards ISO 14000. This is puzzling because: (1) both firms had ISO 9001 quality certification and therefore had experience with implementing other ISO standards; (2) they had well-established environmental management systems; and (3) economic globalization, and expanding outside the US was important to their respective corporate agendas and ISO 14000 could remove a potential hurdle to tap foreign, especially European, markets. Baxter's resistance to ISO 14000 is especially puzzling since it was (and still is) committed to having a state-of-the-art environmental program.

This book attributes the non-adoption of this policy to the absence of either power-based or policy-based processes to work in these firms. Policy supporters could not capture the top management on this policy issue, it did not have the hierarchical power to impose these policies on the skeptics, and it could not muster the ability to induce consensus over these policies. This was due to factors both external and internal to firms. Regarding the former, less hostile external regulatory and legislative environments for the firms in the mid 1990s and the lukewarm support of the EPA to ISO 14000 reduced the incentives for policy skeptics to reassess their opposition or the top management to impose this policy. Internal factors include the emergence of a "green-wall" (see below) in these firms that made their managers less enthusiastic about Type 2 policies, and the firms'(then) extant investments in well-functioning environmental management systems that reduced the incremental benefits of adopting ISO 14000.

Founded in 1946, the ISO is a Geneva-based non-governmental organization that promotes the development and implementation of voluntary international standards. Since the mid 1990s, this organization has been developing a series of environmental management systems standards – ISO 14000 – for manufacturing as well as service organizations. ISO 14000 attempts to replicate the success of ISo 9000 quality control and quality assurance standards in environmental management that were introduced in 1987.[16] ISO 14000 neither replaces industrial codes of practices such as the CMA's Responsible Care nor local laws and regulations. In fact, the latter are treated as minimum requirements

[16] The ideas embedded in ISO 9000 are not new or radical. They share commonalities with Deming's fourteen points (1982, 1992), Total Quality Management, and Malcolm Baldrige National Quality Award criteria.

Table 4.13. *ISO 14000 series: an overview*

ISO series	Description
ISO 14001	Environmental Management Systems: Specifications with Guidance for their Application
ISO 14004	Environmental Management Systems: General Guidelines on Principles, Systems, and Supporting Techniques
ISO 14010	General Principles of Environmental Auditing
ISO 14011	Audit Procedures
ISO 14012	Qualifications Criteria for Environmental Auditors
ISO 14024	Environmental Labeling
ISO 14031	Environmental Performance Evaluation
ISO 14040	Guidelines on Life Cycle Assessment
ISO 14050	Terms and Definitions

Source: Adapted from Puri (1996: 18).

for designing ISO 14000 systems. As shown in Table 4.13, the ISO 14000 series consists of one mandatory standard (ISO 14001) and several non-mandatory standards. ISO 14001 being a compliance standard needs certification by an external auditor: any facility seeking ISO 14000 certification is required to demonstrate that its environmental management systems meet the criteria specified in ISO 14001. In contrast, other ISO 14000 standards are only guidelines and do not require certification.

Baxter and Lilly: response

During the period of this study (mid 1975–mid 1996) both Baxter and Lilly decided against becoming early adopters of ISO 14000. They adopted a wait-and-see attitude and neither of them mandated that their facilities should adopt ISO 14000. This is not to say that they rejected the possibility that their facilities would eventually be ISO 14000 certified. In fact, these firms became "ISO ready" so that, if required, they could have the ISO certification at short notice. Towards this end, they undertook a "gap-analysis" by comparing their environmental systems with the ISO 14000 requirements and identifying areas where their systems needed modification (appendix 4.4). Further (beyond the period of this study), some of Baxter's facilities received ISO 14000 certification. As of July 29, 1997, seven of its 120 facilities are ISO 14000 certified. Lilly's Ireland facility received the ISO 14000 certification in late 1997. Baxter has also commissioned pilots on ISO 14000; five of its facilities will receive the

ISO certification and Baxter will use their experience to evaluate future courses of action (Baxter 1995a). Nevertheless, neither of these firms mandated that all of their facilities should have the ISO certification within a given period of time. This is an interesting puzzle given that these firms have been in the forefront to adopt beyond-compliance policies and have committed millions of dollars to Type 2 policies.

Policy dynamics and analysis

ISO 14000 supporters in Lilly and Baxter could not (and cannot) justify it on the basis of standard investment appraisal procedures such as capital budgeting. They did not have the power to mandate its adoption or the persuasive abilities to convince policy skeptics about its merits. The question then is: unlike other cases examined in this book, why did the power and/or leadership-based processes not succeed for ISO 14000?

To understand Baxter's and Lilly's responses, it is instructive to examine the costs and benefits to individual firms of adopting ISO 14000. Firms face three kinds of costs. First, they bear the costs of creating, adopting, and implementing new management systems. Since ISO 14000 requires extensive documentation many firms may not have the resources for meeting its requirements. Second, firms bear the costs of third-party certification. These costs are significant: typically around $20,000 per facility per audit (Baxter 1996a). Thus, for a firm such as Baxter that had over 200 facilities, third-party certification may cost up to $4 million.[17] Third, firms that invest significantly in research and development are apprehensive about giving outsiders (that is, the third-party certifiers or auditors) access to confidential information. Consequently, to guard against industrial espionage, such firms will need to expend resources for establishing internal protocols on information disclosure.

Firms also face uncertainty on the responsibilities of third-party certifiers. It is not clear if the findings of such certifiers are to be made public and whether the certifiers are required to report instances of environmental violations to the regulators. If the latter is true, it creates disincentives for managers to share compliance information with certifiers. This is important because under many laws, civil and criminal charges can be brought against individual managers for non-compliance with environmental laws. Further, as discussed in the section on environmental audits, the audit findings may not be protected by the attorney–client

[17] Another study estimates that for a firm with twelve facilities, overhead and certification expenses could amount to $1 million per annum; certification expenses alone to be around $25,000 per facility (NSF International 1996).

privilege. This again creates disincentives for US-based firms to invite external certifiers to audit their environmental programs.

ISO 14000 standards also create multiple benefits for firms. ISO 14000 is a key policy response by business firms to environmental issues (Cascio 1994). Firms as a group receive two categories of benefits. First, ISO 14000 replaces multiple environmental standards with a single standard. By increasing transaction costs, multiple standards act as non-trade barriers to international trade. In the 1990s many national and regional environmental standards such as the UK's British Standard 7750, the Canadian Standards Association's CS AZ750–94, and the European Community's Eco-label and Eco-Management and Audit Standards have been proposed. Since these standards have not been fully harmonized, they impede trade. By superseding these standards, ISO 14000 will reduce trade barriers.

ISO 14000 could preempt industry-unfriendly standards made by bureaucrats, thereby reducing the scope of industry-hostile monitoring and enforcement by the regulators and environmental groups. The adversarial relationship between business and government in the United States, especially on environmental issues, is well documented. Though this "adversary economy" co-exists with numerous instances of "capture" (Stigler 1971) of governmental instrumentalities by business firms (Kolko 1963; Bernstein 1955), nevertheless firms generally view government-sponsored standards as being inefficient and "industry unfriendly" (Cascio 1994; International Organization for Standardization 1995a, 1995b). This is because such standards are written by bureaucrats seeking to demonstrate to their constituencies (politicians as well as environmental groups) that they are being "tough with polluters." Many managers believe that firms could deliver comparable levels of environmental performance at a lower cost if they were themselves to write such standards. ISO 14000 provides opportunities for firms to participate in the writing of efficient and easy-to-implement environmental standards.

The ISO 14000 certification process potentially provides credible third-party certification of a firm's environmental management systems. Many managers perceive that certifiers often have a problem-solving attitude. Contrast this with the managerial perceptions of the attitudes of state and federal environmental regulators. For example, inspectors of environmental regulatory agencies are prohibited to give constructive feedback or written comments to firms on the positive aspects of their environmental programs. This has been repeatedly pointed out during my interaction with many firms, including Baxter and Lilly. Similarly, ISO 14000 may reduce public demand for new and more stringent environmental regulations. A widespread acceptance of ISO 14000 may mollify demands of some environmentalists for uniform process standards, which are currently disallowed by the World Trade Organization (formerly, the General Agreement on Tariffs and Trade).

Although significant benefits may accrue to firms as a group, individual firms may have incentives to free-ride and not to invest in adopting ISO 14000 standards. As discussed in chapter 2, an important issue then is whether the institutional design of ISO 14000 creates sufficient net *excludable* benefits for firms to adopt ISO 14000. Further, are these benefits quantifiable, enabling this project to meet the criteria of formal project assessment procedures? The excludable benefits are the following.

First, firms with ISO 14000 certification will adhere to and register for only one standard and not multiple national standards. This will reduce the transactions costs of multiple certifications and enable firms to tap economies of scale. Second, as suggested earlier, the European community may require firms to get ISO 14000 certification to qualify for governmental purchases (Johnson 1994). Hence ISO 14000 certification provides excludable and quantifiable benefits for firms with a business presence in Europe. Note that a similar debate took place in the United States on ISO 9000. Since the European Community had adopted ISO 9000 (and called it EN 29000), it was feared that this may serve as a non-trade barrier for US exporters. Therefore, US firms wanting access to European markets have incentives to get the ISO 9000 certification. For firms catering predominantly to non-European markets, however, ISO 14000 will bring little excludable and quantifiable benefits. Third, firms may have opportunities to reduce costs by improving environmental performance. Firms may lower their insurance costs and have easier and cheaper access to credit and loans by documenting sound environmental practices (Schmidheiny and Zorraquin 1996). Fourth, having ISO 14000 certification demonstrates the intent of a firm to follow environmental laws and regulations in a systematic manner. Even though the EPA does not grant attorney–client privileges, other environmental regulatory agencies may take a less harsh view of minor violations by firms that have such well-established environmental management systems. Fifth, since ISO 14000 may serve as a blueprint for future environmental laws and regulations, its certification may prepare firms for implementing future laws. Finally, ISO 14000 may equip firms for participating in law-making processes. Often, regulatory agencies invite firms, citizen groups, and industry associations to comment on the proposed drafts of a new environmental law or regulation. For example, an internal document of Lilly reported that:

Further gains have been made in our efforts for a proactive involvement in legislative and regulatory issues at the federal and state level . . . Lilly has been particularly active in . . . the EPA's Clean Air Act Advisory Committee. We played a major role during 1992 in developing the proposed regulations for implementing Title V (Operating Permits) under the 1990 Clean Air Act.

At the state level, we have been a major player in the most significant emerging environmental legislations. The principal focus has been on enabling legislations for state implementation of the Clean Air Act. Other activities have centered on the subjects such as pollution prevention, the state's proposal for "good character" requirements for permit pursuance, rulemaking process, and adequate staffing of the state's environmental agency. Our regulatory focus has been on air rules dealing with interim construction permits and the definition of construction and modification. We continue to assist in the development of the "Voluntary Remediation Program." (1992b: 28)

Since the inputs of firms with prior experience with environmental management systems will be more credible, such firms may have opportunities to leverage their expert knowledge acquired during ISO 14000 certification to shape laws and regulations to their advantage.

Then why did Lilly and Baxter adopt a wait-and-see attitude on ISO 14000? Why did power-based dynamics not work? Or, why did some organizational entrepreneurs not take a lead in convincing their skeptical colleagues that having ISO 14000 certification has long-term net benefits? I attribute the absence of power-based and leadership-based dynamics to factors internal and external to firms: the presence of "green-walls" (explained below), extant well-functioning environmental management systems, attorney–client issues, and declining external pressures to "go-green."

There is an emerging literature that suggests that since the mid 1990s "green-walls" have emerged in firms that discourage managers from initiating Type 2 policies. Green-walls manifest in budget and headcount reductions of environmental departments and a reduced representation of senior managers on environmental committees. An article published in *Perspective*, Arthur D. Little's influential management journal, notes that:

A close inspection of today's management of corporate environmental matters indicates that everything is not going smoothly. Although some companies are moving ahead with reliable strategic environmental initiatives, still others are backing away from the broad programs they adopted over the last decade. Why? Because they have hit the Green Wall – they have reached that point at which management refuses to move forward with its strategic environmental program. . . . Early symptoms include . . . deferred decisions because of reduced management support, and an inability to demonstrate return on investment in environmental programs. (1995: 2)

Why has the so-called green-wall emerged only in the mid 1990s and not earlier? How does it impact the adoption or non-adoption of ISO 14000? The green-wall has emerged due to the following factors. First, environmental programs have tended to become victims of their own success. As described in chapter 3, in the late 1970s and the 1980s, firms such as Baxter and Lilly invested significant resources in beyond-compliance environmental programs due to the perceived hostility of the regulators as well as pressures from stakeholders. Cheerleaders, both within and outside firms, proclaiming "win–win" scenarios over-promised benefits of such investments (Walley and Whitehead 1994). Consequently, many environmental programs did not generate expected levels of benefits causing disappointment among many managers. This was also a time when managers were under tremendous pressure to cut costs (see below). Not surprisingly, in the 1990s many managers began closely scrutinizing environmental projects for their impact on quantifiable profits. Supporters of Type 2 policies are, therefore, finding it difficult to convince their skeptical colleagues of the future benefits of such programs.

Second, some environmental managers have adopted cultures that are out of sync with their firm's culture. As one of Baxter's managers put it, "we did not talk the language of business in terms of profits and dollars." In the 1980s, riding on a green-wave, environmental managers felt little need to communicate with managers in other functional areas. Thus the green-wall in the 1990s represents, in part, a backlash from non-environmental managers.

Factors external to firms have also contributed to the emergence of a green-wall. As evident in the initiatives of 104th and 105th US Congresses, there is a political backlash to the perceived excesses of some environmental regulations. The 104th and 105th US Congresses sought to dilute many environmental laws and even cut the budget of the EPA. Though they did not succeed in many of such quests, firms are nevertheless feeling less pressured to gain external credibility by investing in Type 2 environmental programs.

The increased activity in the merger and acquisition markets in the 1990s, from $400 billion in 1990 to about $1.6 trillion in 1997 (UNCTAD 1998), and the perceived pressures of the emerging global economy have made firms very cost conscious. An increased recourse to corporate downsizing for boosting corporate profits has often resulted in the axing of softer areas such as environmental management. This issue has been repeatedly emphasized by managers in Lilly and Baxter. Thus, environmental mangers are increasingly being challenged to make a business case based on quantifiable profits for their beyond-compliance programs. This is not to say that the green-wall encourages firms to violate environmental laws; it only impedes the adoption of Type 2 policies.

ISO 14000 has been critically hurt both by the emerging green-wall and external factors. First, it was initiated in an era when environmental programs are under attack both within and outside firms. Consequently, firms are feeling less pressured to adopt Type 2 policies such as ISO 14000. As I will discuss in chapter 5, external threats are often critical for internal coalitions to develop for championing Type 2 beyond-compliance policies. Skeptical managers are not dismissing ISO 14000 altogether; they are recommending a wait-and-watch policy. The issue is not whether or not ISO 14000 is beneficial in the long run; there is a fair degree of agreement that it will have benefits in the long run. Rather, the debate is when a firm should invest in ISO 14000 certification. ISO 14000 supporters are finding it difficult to convince skeptics that it is worthwhile to be an early adopter.

Second, ISO 14000 offers little value to firms that already have well-established environmental management systems. Such path-dependencies that are emphasized by new institutionalists – past policies influence current and future outcomes – have reduced the incremental benefits of ISO 14000. Many firms already subscribe to some well-recognized industrial codes of practice such as Responsible Care. This again reduces their need to gain external legitimacy by joining another industrial code of practice such as ISO 14000.

This discussion summarizes the mood within Lilly and Baxter on ISO 14000. Lilly's resistance to ISO 14000 can be traced to the factors discussed above. First, Lilly already had a well-established environmental control process: Environmental Quality System (EQS). This system is modeled on the lines of stringent product quality systems employed by pharmaceutical companies for meeting the US Food and Drug Administration's strict guidelines. At the heart of EQS is a computer-based Plant Site Environmental Compliance Listing (PECL) system (appendix 4.2). Lilly implemented this system in 1994. PECL simplifies environmental permits and regulations into an easy-to-use plant site compliance list. As a result, every tank, every pump, and every compliance point is identified for its environmental requisites (Lilly 1995a).

Second, as discussed previously, Lilly is actively implementing a widely accepted code of industrial practice: the CMA's Responsible Care program. Lilly has very active organizational involvement in the CMA with John Wilkins of Corporate Environmental Affairs chairing the Indiana-Ohio-Kentucky chapter of Responsible Care. Although Responsible Care and ISO 14000 have many commonalities, they also

Table 4.14. *Lilly's environmental program staff*

Year	Full-time equivalent in person years	Person years/net sales in $billions
1990	218	52.2
1991	271	59.8
1992	313	63.1
1993	340	65.4
1994	338	59.2
1995	334	49.4

Sources: (1) Lilly (1994a: 31).
(2) Lilly (1997a).
(3) Lilly (1992b:10).
(4) Lilly (1995c:34).

differ on some issues. Unlike ISO 14000, Responsible Care requires that firms undertake community-outreach programs. On the other hand, ISO 14000 requires external certification where as Responsible Care does not. Thus, apart from the costs of getting the ISO 14000 certification, there were additional costs involved for Lilly to modify its management systems and adopt ISO 14000 compatible systems. However, Eli Lilly would have received little incremental external validation for its environmental program by having its facilities ISO 14000 certified.

Third, some sort of a green-wall had developed in Eli Lilly. As shown in table 4.14, since 1993, Lilly's human resource commitment to environmental programs as a proportion of its net sales has declined. Further, as discussed in chapter 3, Eli Lilly's corporate environmental program, which was previously headed by a manager of the rank of Vice-President, is now headed by a manager of the rank of a Director. This again reflects the diminished clout of environmental affairs within the organization. As discussed previously, the emergence of a green-wall is a symptom of a more close-fisted approach to Type 2 policies such as ISO 14000.

Fourth, Eli Lilly has had little experience with external audits of its environmental program. Since ISO 14000 requires third-party certification, some managers were apprehensive about the potential misuse of audit information, especially since the EPA does not grant attorney–client privilege on environmental audits.

Baxter also adopted a wait-and-see policy on ISO 14000. It did not mandate its facilities to seek the ISO 14000 certification. Verie Sandborg (1996), Baxter's Environmental Affairs Manager, emphasized that:

We are going to take a harder look at the requirements contained in the standard and what value ISO 14001 can add to our program . . . We essentially have all the basic components. On the whole, our program is comparable [to ISO 14001]. (1996: 23)

Like Lilly, Baxter has well-established environmental management systems. As discussed in the cases on environmental audits, Baxter regularly invites Arthur D. Little to reassess its state-of-the-art standards and audit its environmental systems against these standards. Therefore, ISO 14000 contributed little to improving the design or working of Baxter's environmental management systems. Further, Baxter's environmental programs had received wide recognition. Consequently, Baxter gained little incremental external credibility by investing in ISO 14000 certification.

A green-wall also impacted Baxter's environmental programs in a subtle way. Throughout the 1990s Baxter had a Vice-President, William Blackburn, heading its corporate environmental affairs. Thus, there was no palpable downgrading of intra-organization clout of the environmental department. Environmental managers also seemed to have preempted a major demand associated with an emerging green-wall: a demand that beyond-compliance projects be justified on business grounds. As discussed in chapter 3, Baxter's Corporate Environmental Affairs calculates the net monetary impact of environmental programs on Baxter's profits. However, this proactive stand on calculating the business impact of policies backfired in the case of ISO 14000; its policy supporters could not make a business case for early adoption. In addition, since there is educed pressure on firms from the external environment for visible environmental initiatives, ISO 14000 policy supporters could not forge a coalition in its support.

To conclude, power-based or leadership-based processes did not work for three reasons. First, these firms had (and have) well-functioning environmental management systems that had wide-spread external credibility. This path dependency made ISO 14000 less attractive to them.[18] Second, neither of these firms faced any external threats that could encourage power dynamics or facilitate emergence of coalitions to push for ISO 14000. Third, both firms faced the emergence of some sort of green-wall. This challenged the adoption of environmental policies, such as ISO 14000, that cannot be justified on the basis of established procedures of investment analysis. Both power-based and leadership-based processes require individuals to take ownership of a given program but the green-wall discouraged managers from championing ISO 14000.

[18] There is a well-established literature on the impact of path dependency on adoption of standards. Important works include Katz and Shapiro (1983, 1985), Bessen and Saloner (1988), and Saloner and Shepard (1991).

Appendix 4.1 Six codes of Responsible Care

Code 1. Community awareness and emergency response

Approved: November 6, 1989
Objective: Ensure emergency preparedness and foster community-right-to-know.
Requirements:
- Develop community outreach programs for communicating information on environment, health, and safety aspects of facility's operations.
- Develop programs for responding to emergencies. Test these programs annually involving all relevant stakeholders.

Code 2. Pollution-prevention

Approved: April 6, 1990 (waste and release practices); September 5, 1991 (waste management practices)
Objective: Promote pollution-prevention and waste-minimization.
Requirements:
- Document waste generation, estimate their releases to various media, and evaluate their potential health, safety, and environmental impacts.
- Seek employee and public input for developing and implementing waste-minimization and pollution-prevention policies.
- Emphasize source-reduction and include pollution-prevention as an objective at the research and development stage.

Code 3. Process safety

Approved: September 10, 1990
Objective: Prevent industrial accidents.
Requirements:
- Develop process safety programs. Document and measure safety performance.
- Audit safety systems. Conduct safety reviews of new-modified facilities before commissioning them.
- Train employees in safety procedures.

Code 4. Distribution

Approved: November 5, 1990
Objective: Minimize risks posed by transportation and storage of chemi-

cals to carriers, customers, contractors, employers, and the environment.

Requirements:

- Evaluate risks associated with existing modes of transportation and distribution.
- Train employees, carriers, and contractors on the regulations and best practices.
- Regularly review the performance and practices of carriers.
- Develop an emergency response plan for dealing with transportation accidents.

Code 5. Employee health and safety

Approved: January 14, 1992
Objective: Protect and promote health and safety of employees and visitors at facilities.

Requirements:

- Develop and review occupational safety systems; audit them; train employees.
- Select vendors and contractors that follow the above guidelines; audit them.
- Investigate trends in workplace illness, injuries, and accidents.

Code 6. Product stewardship

Approved: April 16, 1992
Objective: Promote safe handling of chemicals from their initial manufacture to their distribution, sale, and disposal.

Requirements:

- Develop a corporate plan on product stewardship.
- Incorporate environmental, health, and safety concerns on the product and process development stage.

Appendix 4.2 Baxter's environmental awards

1988–1993

BAXTER OPERATIONS received 37 environmental awards.

1994

BAXTER
The Issue Exchange
W. Howard Chase Award for Excellence in Issue Management

CAROLINA, PUERTO RICO
Environmental Quality Board
Best Air Program Award

CAROLINA, PUERTO RICO
Puerto Rico Aqueduct and Sewer Authority
Water Quality Award

CARTAGO, COSTA RICA
Chamber of Industries of Costa Rica and the Municipality of San Jose
Environmental Conservation Award

CARTAGO, COSTA RICA
Ministry of Natural Resources
Environmental Flag

CASTLEBAR/SWINFORD, IRELAND
Minister for Environmental Protection, Ireland
Good Environmental Management Award

CASTLEBAR/SWINFORD, IRELAND
European Better Environment Awards for Industry
Commendation Award in the Good Environmental Management Category

ENVISION™ RECYCLING PROGRAM
I. V. Systems Division
Society of Plastics Engineers
First Annual Award for Recycling

HAYWARD, CALIFORNIA
City of Hayward
Earth Day Award

HAYWARD, CALIFORNIA
California Integrated Waste Management Board (CIWMB)
Waste Reduction Award

I. V. SYSTEMS
Illinois Environmental Protection Agency
Star Partner Award

IRVINE, CALIFORNIA (CVG)
City of Irvine
Environmental Excellence Award

IRVINE, CALIFORNIA (CVG)
California Integrated Waste Management Board (CIWMB)
Waste Reduction Award

IRVINE, CALIFORNIA (CVG)
National Awards Council for Environmental Sustainability
Environmental Achievement Certificate

MARICAO, PUERTO RICO
Puerto Rico Solid Waste Authority
Recycling Program Award

MONCTON, NEW BRUNSWICK, CANADA
Greater Moncton Chamber of Commerce
Park Beautification Award

MOUNTAIN HOME, ARKANSAS
Arkansas Environmental Federation
Hazardous Waste Minimization Award

MOUNTAIN HOME, ARKANSAS
Arkansas Recycling Coalition
Corporate Recycler of the Year

NORTH COVE, NORTH CAROLINA
North Carolina Recycling Association
Merit Award

OAKLAND, CALIFORNIA
CIWMB
Waste Reduction Award

POINTE CLAIRE, QUEBEC, CANADA
West Island Chamber of Commerce (Montreal)
Award for Environmental Achievement

RIVERSIDE, CALIFORNIA
Riverside County
Recycling Award

THETFORD, ENGLAND
Beazer Homes
Environment Award

WINNIPEG, MANITOBA, CANADA
Manitoba Round Table of Environment & Economy
Sustainable Development Award of Excellence

1995

ALLISTON, ONTARIO, CANADA
Baxter Canada
Regional Award for Pollution Prevention

AÑUSCO, PUERTO RICO
Puerto Rico Solid Waste Management Authority
Recycling

BAXTER
Lake County Youth Conservation Corps
Community, Careers, Conservation Award

BAXTER
Illinois Parks and Recreation Association and Illinois Association of Park Districts
Community Service Award for the Advancement of Parks

CARTAGO, COSTA RICA
Ministry of Natural Resources
Ecological Flag

HAYWARD, CALIFORNIA
California Integrated Waste Management Board (CIWMB)
Waste Reduction Award

IRVINE, CALIFORNIA (CVG)
City of Irvine
Environmental Community Outreach Award

IRVINE, CALIFORNIA (CVG)
City of Irvine
Water Conservation Award

IRVINE, CALIFORNIA (CVG)
Mayor of Irvine
Waste Reduction Commendation

IRVINE, CALIFORNIA (CVG)
Mayor of Irvine
Arbor Day Tree Planting Commendation

IRVINE, CALIFORNIA (CVG)
Mayor of Irvine
Environmental Community Outreach Award

IRVINE, CALIFORNIA (CVG)
Harbor, Parks and Beaches Association
Park Cleanup Recognition

IRVINE, CALIFORNIA (CVG)
California State Legislature
Environmental Award

IRVINE, CALIFORNIA (CVG)
U.S. Congressmen
Environmental Community Outreach Recognition

IRVINE, CALIFORNIA (CVG)
Renew America
Environmental Sustainability Award

IRVINE, CALIFORNIA (CVG)
California Integrated Waste Management Board (CIWMB)
Waste Reduction Award

LOS ANGELES, CALIFORNIA
Catalina Harbor Clean-up
Golden Flipper Award

MCGAW PARK, ILLINOIS
Lake County Solid Waste Board
Award for Environmental Education

MISSISSAUGA, ONTARIO, CANADA
City of Mississauga
Best Environmental Program Award

MISSISSAUGA, ONTARIO, CANADA
Region of Peel
Outstanding Industrial and Commercial Environmental Initiatives and Waste Reductions

MOUNT PEARL, NEWFOUNDLAND, CANADA
Government of Newfoundland
Environmental Award

MOUNTAIN HOME, ARKANSAS
State of Arkansas
Pollution Prevention Award

VERIE SANDBORG
National Association for Environmental Management
Certificate of Recognition

CURTIS STEPHAN
National Association for Environmental Management
Environmental Excellence Award

1996

AIBONITO, PUERTO RICO
Solid Waste Authority of Puerto Rico
Recycling Program Award

ALLISTON, ONTARIO, CANADA
Recycling Council of Ontario
Waste Minimization Award

ALLISTON, ONTARIO, CANADA
Arbor Committee of New Tecumseth
Corporate Green Challenge Winner

BAXTER
Lake County Youth Conservation Corps
Support Recognition

BAXTER
U.S. Environmental Protection Agency
Certificate of Appreciation for Voluntary Air Toxics Reduction

BAXTER
U.S. Environmental Protection Agency
WasteWi$e Sustained Leadership in Waste Prevention Recognition

CARTAGO, COSTA RICA
Ministry of Natural Resources
Ecological Flag AA (2-Year Winner)

EDMONTON, ALBERTA, CANADA
Pollution Probe Foundation
Commuter Challenge Most Participation Award

ENVIRONMENTAL PERFORMANCE REPORT
Chicago Chapter of the International Association of Business Communicators
Spectra Award of Excellence

ENVIRONMENTAL PERFORMANCE REPORT
District 4 of the International Association of Business Communicators
Silver Quill Award of Excellence

ENVIRONMENTAL PERFORMANCE REPORT
Healthcare Marketing Report
Healthcare Advertising Merit Award

RAFAEL GUZMAN
Santiago de Cali University
Distinguished University Teacher (for teaching environmental mangement at the university)

HARVARD, CALIFORNIA
California Integrated Waste Management Board (CIWMB)
Waste Reduction Award

ROUND LAKE, ILLINOIS
(I. V. SYSTEMS DRUG DELIVERY)
Governor of Illinois
Pollution Prevention Award

IRVINE, CALIFORNIA (CVG)
CIWMB
Waste Reduction Awards Program (4-Year Winner)

IRVINE, CALIFORNIA (CVG)
Renew America
Environmental Sustainability Award

LESSINES, BELGIUM
Belgium Federation of Chemical Industry
Responsible Care Special Recognition

MARSA, MALTA
US Environmental Protection Agency
Stratosphere Ozone Protection Award

MCGAW PARK, ILLINOIS
National Association for Environmental Management
Special Recognition Award for Environmental Excellence

MISSISSAUGA, ONTARIO, CANADA
Mississauga Board of Trade
Environmental User Award

THOUSAND OAKS, CALIFORNIA
City of Thousand Oaks
WasteWatch Gold Award for Outstanding Achievements in Recycling

1997

BAXTER
Lake County Youth Conservation Corps
Recognition of Contributions

BAXTER
National Association of Physicians for the Environment
Inaugral Award for Business Activity to Protect Our Environment and Health

CAROLINA, PUERTO RICO
Solid Waste Authority
Recycling Award

IRVINE, CALIFORNIA (CVG)
City of Irvine
Earth Day Environmental Outreach Award

IRVINE, CALIFORNIA (CVG)
RENEW America
Environmental Sustainability Award

ROUND LAKE, ILLINOIS (I. V. SYSTEMS)
Solid Waste Agency of Lake County
4R Award (for Waste Reduction)

Source: Baxter (1997b).

5 Beyond-compliance: findings and conclusions

This book examined environmental policymaking within two firms – Baxter International Inc. and Eli Lilly and Company – by exploring the internal processes that led them to selectively adopt Type 2 policies. Ten cases were examined: four common to these firms (underground tanks, 33/50, ISO 14000, and environmental audits), and one each specific to them (Responsible Care to Lilly and Green Products to Baxter).[1] All cases pertain to policymaking during 1975 to mid 1996. Cases were selected to ensure variation on independent variables (factors external and internal to firms). Further, as advised by King, Keohane, and Verba (1994), given the small sample size, I also ensured variation on the dependent variable (policy adoption or non-adoption). This discussion is summarized in table 5.1.

Firms' environmental policies were classified on two attributes: (1) whether they meet or exceed the requirements of laws and regulations, and (2) whether or not they meet or exceed the criteria specified in investment appraisal procedures. Based on this classification, four modal types of environmental policies were identified: Type 1 (those which go beyond-compliance and also meet or exceed the profit criteria), Type 2 (those which go beyond-compliance but cannot or do not meet the profit criteria), Type 3 (those which are required by law and also meet or exceed the profit criteria) and Type 4 (those which are required by law but cannot or do not meet the profit criteria).

Stringent monitoring and enforcement of environmental laws in industrialized countries, particularly the US, has ensured that managers have few incentives to systemically violate these laws. Consequently, the book focused only on beyond-compliance policies examining why firms selectively adopt Type 2 policies. The neoclassical economic theory suggests that firms' policies directly correspond to external stimuli, whether in the form of governmental regulations or market signals. It predicts that firms

[1] In this chapter, internal and external audits as well as Phase I and Phase II of Responsible Care are analyzed separately.

Table 5.1. *Findings*

	Baxter	Lilly
Underground tanks	Adopted: leadership	Adopted: leadership
EPA's 33/50	Adopted: leadership	Adopted: leadership
Responsible Care	n.a.	Initially not adopted; Finally adopted: leadership
ISO 14000	Not adopted	Not adopted
Internal audits	Adopted: power	Adopted: leadership
External audits	Adopted: power	Not adopted
Green products	Adopted: leadership	n.a.

would only adopt those policies that are either required by laws and regulations (Type 3 and Type 4) or that can be demonstrated *ex ante* as being profitable (Type 1).

The neoclassical theory is less useful in explaining why firms selectively adopt Type 2 policies. Sociological institutional theory and stakeholder theory are also inadequate to explain the research puzzle, although they contribute to understanding why firms adopt policies that may not deliver short-term or even long-term quantifiable profits. This is because focusing *solely* on how external factors shape firms' incentives (as the above theories do), *though necessary, is insufficient* to explain this puzzle. For example, only 13 percent of the eligible US firms adopted the Environmental Protection Agency's (EPA's) 33/50 program (chapter 4). Even among the top 600 companies, only 64 percent joined this program (Sarokin 1999). Any theory or perspective that focuses on the coercive character of the EPA (an external factor) as the sole explanatory variable is, therefore, under-specified to explain variations in firms' responses to the 33/50 program. Being under-specified means that as a part of its causal explanation, the neoclassical theory leaves out one or more of the important independent variables; in the context of beyond-compliance policies, internal processes of firms constitute the omitted variables.

Given our inability to unbundle firms by employing the neoclassical *theory*, the institutional *theory*, and the stakeholder *theory*, I turned to the new-institutionalist *framework*.[2] This was also an ontological departure since managers within firms, and not firms themselves, became my unit of analysis. Both new-institutionalists and neoclassicalists assume methodological individualism. However, by treating firms as unitary actors, neoclassicalists make no distinction between firms and individual actors who

[2] The distinction between a theory and a framework is discussed subsequently.

work within these firms. Since many scholars of organizational theory (including those subscribing to the behavioral theory of the firm) view firms as composite actors, a strategy of unpacking firms is not unique to new institutionalists.

Next, I explained the adoption or non-adoption of Type 2 policies based on preferences (policy supporters versus policy skeptics) and endowments of actors (specifically, their position in a firm's hierarchy) as articulated within: (1) internal institutions and structures of their firm, and (2) the external environment in which their firm functioned. My strategy therefore had three steps. First, I opened up the "black-box" called the firm to identify the main actors in the firm's environmental policymaking, their preferences for Type 2 policies, their position in the hierarchy, and the firm's internal procedures and structures. Second, I linked actors, procedures, and structures to processes of environmental policymaking. Processes imply two things: (a) the criteria employed by managers for assessing the profitability of projects, including environmental projects and (b) the way in which managers interpret these criteria to support their desired projects, and, as a result, how projects may not always be subjected to such formal appraisal procedures. Third, since firms constantly negotiate with the external world, I examined how policy supporters, successfully or unsuccessfully, invoked such external factors to influence internal processes.

New institutionalists treat policies of firms as institutions, and firms themselves as organizations. Hence the book examined the adoption or non-adoption of a specific category of institutions: Type 2 policies. New institutionalists focus on two sets of questions: first, how do institutions evolve in response to individual preferences, strategies, and endowments, and, second, how do they affect policy outcomes? This study focused on the first question by examining the adoption and non-adoption of a firm-level institution: Type 2 policies.

New institutionalists assume methodological individualism, bounded rationality, and pursuit of self-interest. Given their bounded rationality, managers employ established procedures such as capital budgeting to assess *ex ante* the profitability of projects, thereby attempting to maximize shareholders' wealth. In capital budgeting, managers project future cash flows emanating from a project and discount them by the firm's cost-of-capital. Cash flow projections are the best managerial estimates of an uncertain future. To convert issues of uncertainty into issues of risk, managers weight the cash flows by their subjective probabilities. They periodically update probabilities in light of market or regulatory changes. Needless to say, managers' assessments of the future vary substantially. To map out variations in projected profit and cash flows, managers subject these estimates to sensitivity analysis. Thus, managers adopt standard operating procedures to cope with uncertainty and to establish transparency and impartiality in the project appraisal process. A project is *ex ante* profitable if it meets the criteria of such established procedures. Importantly, *ex ante* profitable projects may result in *ex post* losses if market or regulatory conditions change in unanticipated ways.

There are other methods of investment analysis as well, such as full-cost accounting, and life-cycle analysis that firms could potentially employ. These methods force firms to take into account long-term social (as well as private) costs, thereby internalizing environmental externalities. I did not examine these alternative investment appraisal techniques for several reasons. First, Lilly and Baxter do not routinely employ any of these methods for appraising projects (while they do employ capital budgeting). Further, even as a non-routine measure, they did not employ them in making decisions on Type 2 policies examined in chapter 4. Second, since the methodologies and operationalization of these techniques is not standardized, they lack legitimacy, especially with the finance and accounting managers. Finally, these techniques may not be compatible with the dominant paradigm that the main objective of firms is to maximize shareholders' wealth, an objective that capital budgeting focuses on.

I argued that certain managers may think that a given Type 2 policy is beneficial for their firm in the long-run even though its impact on profits cannot be quantified. Clearly, such policies cannot satisfy the requirements of established procedures of investment analysis. These managers then face a difficult task of convincing skeptics that this policy should not be subjected to such procedures. In these situations the internal policy-making dynamics become important. Policy supporters *do not always* succeed (as in ISO 14000 and external audits in Lilly). If they do, it reflects power-based or leadership-based processes.

Power-based theories view firms as representing the domination of one set of actors over another. They, therefore, predict that Type 2 policies will be adopted only if they have powerful sponsors such as hierarchically superior managers.[3] Some of Baxter's managers opposed inviting Arthur D. Little (ADL) to help them define state-of-the-art environmental standards and to audit Baxter's environmental programs against these standards (chapter 4). Interventions by very senior managers including

[3] Theoretically, there is another policy category: Type 1 policies that were not adopted due to the opposition from powerful actors. This theoretical category has little practical relevance since senior managers, whose compensation is often linked to quantifiable profits and the firm's stock price, will have few incentives to oppose such policies. I could not find instances of such opposition in Baxter or Lilly.

Baxter's Chief Executive Officer, ensured that such objections were overruled. Consequently, this policy was adopted by imposition.

Leadership-based theories suggest that power-based (and transaction cost) explanations cannot fully explain the nature of firms. They argue that firms emerge only through the intervention of leaders who can convince other managers to reassess their assumptions and preferences regarding the costs and benefits of collective action. Policies carry leaders' imprints in that firms would not have adopted them in the absence of leaders' interventions. Type 2 policies on underground tanks, the 33/50 program, and Responsible Care were initially opposed by some managers since they do not meet the formal profit criteria. Over time, however, policy supporters succeed in convincing policy skeptics that these policies indeed serve the long-term interests of their firms, although their profit contributions cannot be quantified. Such policies are eventually adopted by inducing consensus.

Ex ante, profits are mere projections. Of course, by considering alternative scenarios (through sensitivity analysis), managers endeavor to make more rigorous projections, a task that has been greatly facilitated by computer packages. In some instances, sole reliance on formal assessment procedures is not useful since it is not possible to quantify the profit impact. This creates a political space for "discursive struggles" within firms and opportunities for managers who claim to have a "vision" to push through their pet projects. These policies are implemented if either such managers are able to convince others that such projects help the firm in the long-run (leadership-based processes) or if they have the authority to impose their "vision" on others (power-based processes). Although the policy supporters often claim increases in long-term profits, they provide no estimates. This suggests that in some instances profit no longer remains an "objective" concept whose assessment is invariant across actors. I am not arguing that established methods of project appraisal are irrelevant. They are indeed relevant and that is why it is difficult for the supporters of Type 2 policies to justify why their pet policy should not be subjected to established project appraisal procedures. Such exceptions do occur, especially in the realm of social policies such as environmental policies. This book attempted to understand the processes that lead to such exceptions and the conditions under which they occur.

To summarize, neoclassical economics predict that managers employ some "objective" criteria to assess the profitability of policies. Profitability would be assessed *ex ante*, and only policies meeting or exceeding this criteria would be adopted. In the context of Type 2 policies, managers cannot or do not employ such well-defined criteria. For example, it is difficult to assess the profitability of environmental audits that do not require up-front capital expenditure and that do not generate revenues (ISO 14000). Even in some circumstances (such as underground storage tanks and the 33/50 program) that involved significant capital expenditures, policy supporters succeeded in not employing established appraisal procedures.

This study has important implications for environmental policy debate. Many argue that firms are often oblivious of the low-hanging fruit – "win–win" projects that generate profits and deliver superior environmental performance (Porter 1991; Porter and van der Linde 1995). Hence, stringent environmental laws may force firms to harvest this low-hanging fruit, thereby benefitting themselves and also the natural environment. Win–win strategies assume that firms can earn quantifiable profits by adopting environmentally progressive policies. Some projects (Type 1 policies) certainly fall in this category, especially the ones that lead to pollution prevention or reduction. Type 2 policies such as adopting new management systems can seldom be demonstrated as being *ex ante* profitable. A focus on win–win projects as a pillar of environmental policy is misplaced. Many "desirable" environmental policies could conceivably benefit firms in the long run only and it is often difficult to justify on traditional economic grounds. The win–win rhetoric creates false expectations that lead to backlash when not met. The emergence of the so-called "green-wall" in the mid 1990s (chapter 4) was primarily due to the managerial skepticism about the economic soundness of environmental policies. Even if low-hanging fruits exist, many of them have been already harvested by firms in response to public pressure, new managerial consciousness, and stringent regulations. Now, firms are struggling to harvest the high-hanging fruits or fruits from some other tree, and these may not constitute a win–win scenario. This book, therefore, argues that the new genre of environmental policies and regulations need to be justified on non-economic grounds as well because their economic rationale may be difficult to demonstrate.

Theoretical implications

Research objectives often fall in the realm of either theory testing or theory building. Theory testing is useful when the existing theoretical tools are sufficient to examine a given puzzle. Of course, in the process of testing existing theories, researchers may suggest modifying them. Theory building is useful when existing theoretical tools are insufficient in explaining a given research question. For example, I classify Coase's (1937) landmark article, "The Nature of the Firm," as being in the realm of theory building since the neoclassical economic theory did not sufficiently explain the major theoretical question: why do firms arise in the first place? This book could also be viewed in the realm of theory building. Existing theories – neoclassical economics, sociological institutional theory, and stakeholder theory – inadequately explain why firms selectively adopt Type 2 policies. I have, therefore, employed two theories that together explain this research question. However, this is only an initial step in developing better theory to understand this question. I have attempted to understand the notion of efficiency by differentiating between substantive efficiency and procedural efficiency, and argued that to examine intra-firm dynamics, procedural efficiency is more appropriate. As a second-round effect, substantive efficiency is predicted to influence procedural efficiency. However, this impact is mediated by a variety of institutions such as competition in the market as well as managerial abilities to learn from such market signals.

This book has three theoretical implications. First, ontologically, it makes an argument in favor of methodological individualism: the policies of composite actors should eventually be traced to preferences, endowments, and strategies of individual actors, as articulated within their internal institutions. Second, many social phenomena such as corporate responses to beyond-compliance policies cannot be adequately understood by employing one theory; rather this requires employing multiple theories and a common framework to link them. Third, it makes an argument for "bringing back leadership" to political economy, since leaders often play crucial roles in institutional evolution and change. I elaborate on these implications below.

Composite actors

The neoclassical theory (as well as sociological institutional and stakeholder theories) treat firms as undifferentiated unitary actors responding only to external factors. This spartan view of the firm serves well to predict firms' responses to Type 3 and Type 4 policies (both are required by law) and Type 1 policies (those that meet or exceed the requirements of capital budgeting) but not Type 2 policies.

How can the selective adoption of Type 2 policies be explained? There are two strategies, both sacrificing parsimony to increase the explanatory power of the theory. The first strategy is to view firms as differentiated unitary actors and not as undifferentiated unitary actors. With this additional assumption, one could argue that differentiation across firms makes them respond differently to the same external stimulus. The second strategy is to treat firms as differentiated composite actors and focus on their internal processes.[4] One could then argue that the differences in internal processes explain variations in response to Type 2 policies. I elaborate on these two strategies below.

In employing the first strategy, firms could be differentiated on the basis of a variety of indicators such as assets and sales (large firms versus small firms) or industry type (engineering firms, chemical firms, electronic firms, etc.). As discussed in chapter 1, UNCTAD's benchmark survey on MNEs' environmental practices suggests that MNEs' environmental programs are a function of three factors: size of the firm, nationality of the parent firm, and industry type (UNCTAD 1993). It reports that larger MNEs (with sales above $4.9 billion) have stronger environmental programs than smaller MNEs.

A strategy of treating firms as differentiated unitary actors is considerably superior to one that treats them as undifferentiated unitary actors. Such differentiation based on their external attributes enhances the ability of theory in explaining why firms with given attributes tend to adopt specific kinds of environmental programs. In effect, by diluting parsimony, our theoretical tools have gained superior explanatory and predictive power.

We still cannot explain why, in response to the same external stimulus, firms sharing similar attributes adopt different policies. For example, while Baxter and Lilly share characteristics identified in the UNCTAD survey, they have responded differently to the demands of environmental groups for transparency in their environmental programs. Baxter invites external auditors to evaluate its environmental programs whereas Lilly does not. Further, we are unable to explain why a given firm responds differently to similar external stimuli. For example, Lilly has responded differently to the adoption of industrial codes of practices; it has adopted Responsible Care but not ISO 14000.[5]

What is the next step? One strategy could be to make the classification of firms' attributes more complex. However, there is no theoretical reason why this additional complexity will improve the theory's explanatory power. I am of the view that any classification must have a theoretical basis: we must have a logic to classify in the manner that we do. Though one could observe patterns and infer theoretical reasons why

[4] There is a third strategy as well: firms as undifferentiated composite actors consisting of employees with identical endowments and preferences and functioning in identical sets of internal institutions. Clearly, this category has little practical relevance. It is also not a theoretically useful category since the objective of unpacking firms is to identify differences in their internal structures and processes, and link these differences to variations in policy outcomes.

[5] For an elaboration of this argument, see Prakash (1999a).

such patterns arise, such explanations are under-specified in that there could be multiple reasons for the same pattern. This book, therefore, does not adopt this approach.

Perhaps, a paradigmatic shift is required to change the focus of inquiry from firms' external factors to their internal ones. This requires employing the second strategy of unbundling firms; treating them as differentiated composite actors, and not as differentiated unitary actors. This calls for revising the ontological assumptions; instead of treating firms as units of analysis, the focus shifts to individual managers within firms as units of analysis.

An examination of inter-manager interactions, particularly between policy supporters and policy skeptics, should explain the following: (1) in response to the same external stimulus (such as stakeholder scrutiny), why some firms adopt certain Type 2 policies, while others do not (only Baxter invites external auditors); (2) in response to similar external stimuli (for example, industrial codes of practice), why the same firm adopts different policies (Lilly adopts Responsible Care but not ISO 14000). Such inter-manager interactions could be dissimilar within and across firms due to factors such as: (1) variations in firms' internal decision-making institutions; (2) levels of opposition of policy skeptics (specifically, the "losers" from changes in collective-choice-level institutions); (3) the hierarchical power of policy supporters in relation to policy skeptics; (4) the credibility and persuasive abilities of policy supporters in rallying policy skeptics; and (5) the managerial perceptions of the coercive power of external institutions that are encouraging or discouraging adoption of a given Type 2 policy.

Multiple theories and common framework

The second theoretical implication of this book is that explaining social phenomena (Type 2 policies in this book) often requires multiple theories linked by a common framework. The distinction between frameworks and theories is emphasized in the literature on the Institutional Analysis and Development framework developed at the Workshop in Political Theory and Policy Analysis (Ostrom 1990; Ostrom, Gardner, and Walker 1994). According to Elinor Ostrom:

> A framework helps to identify the elements and relationships among these elements that one needs to consider for institutional analysis . . . they organize diagnostic and prescriptive enquiry . . . provide a metatheoretical language to compare theories . . . attempt to identify universal elements that any theory relevant to the same kind of phenomena would need to include. (1996: 4)

In contrast, theories help analysts to specify elements of a framework that are relevant to a given question. They make specific assumptions to diagnose this question, explain its sources, and formulate some solutions. Several theories are usually compatible with any framework (Ostrom 1996). A given theory may be compatible with other frameworks as well. For example, power theories are compatible with the Marxian framework. This book employs two sets of theories linked together within a new-institutionalist framework.[6] Both these theories view institutions as potential artifacts. However, they differ in their explanations of how and why institutions emerge. No theory alone can explain all aspects of my research question.

Neoclassical economics predict that Type 2 policies would not be adopted. Power-based theories predict that Type 2 policies would be adopted if policy supporters have sufficient coercive power to over-rule policy skeptics. Thus, such policies may be adopted without changes in dissenters' preferences. I disagree with scholars who suggest that power-based explanations are preeminent, and leadership is only a subtle exercise of power. Later in the chapter, along with this issue, I discuss my strategy to infer whether or not policy skeptics have changed their preferences.

In contrast to power-based processes, leadership-based policies involve changes in preferences. This requires an assessment of preference intensity. My strategy has been to examine the extent of organizational change, thereby assessing the perceived "loss" of "losers" from such changes. If the loss is significant, losers can be predicted to exhibit strong preferences for not adopting Type 2 policies, and would therefore have few incentives to revise their preferences. As a result, if such policies are implemented at all, it would be through power-based processes.

Leadership-based policies bear the imprints of particular leaders who employ their persuasive skills to convince policy skeptics of the long-term payoffs of adopting Type 2 policies. Over time, in some instances, policy skeptics change their preferences, and such policies are then adopted consensually.

Do these theories constitute alternative explanations; that is, do they suggest different logics for understanding a given phenomenon? If so, do we need to prioritize these theories? Prioritizing can have two meanings.

[6] In the International Business literature, Dunning (1993) employs the eclectic paradigm – the OLI framework – to link three sets of theories that together explain why firms invest abroad. These theories focus on three sets of variables that pertain to the competencies of *organizations*, the advantages for firms to *locate* abroad, and the need to *internalize* transactions within the firm instead of using market processes.

First, one could rank these theories on their explanatory and predictive powers. This is not done since I see them as working *together* to explain a broader puzzle. Second, one could prioritize these theories in terms of the research chronology. This study followed a particular format in my analysis of the ten cases. First, it employed neoclassical economics for understanding the adoption or non-adoption of a particular policy. This is because it is generally assumed that the preeminent objective of firms is pursuit of profits. Since neoclassical economics could not explain the selective adoption of Type 2 policies, the study employed leadership-based and power-based explanations.

Since both power-based and leadership-based perspectives add to our understanding of social phenomena, why should there be an insistence that one perspective must always prevail?[7] Why not view these perspectives as complementary that together help us to understand complex social reality? In employing three models (rational actor, organizational process, and governmental politics) to examine the central puzzles of the Cuban missile crisis, Allison notes:

> The three models are obviously not exclusive alternatives. Indeed, the paradigms highlight the partial emphasis of each framework – what each magnifies and what it leaves out. Each concentrates on one class of variables, in effect, relegating other important factors to a *ceteris paribus* clause. The models can therefore be understood as building blocks in a larger model of the determinants of outcomes. (1971: 275)

The distinction between a framework and a theory provides a persuasive rationale for employing multiple theories. Theories and frameworks constitute two different levels of theoretical inquiry. Often by focusing on the incorrect level of analysis (the level of theory) we find ourselves in midst of unnecessary contestations on the superiority of one theory over another. An understanding of why firms adopt or do not adopt beyond-compliance policies would have been incomplete had I employed only leadership-based or power-based theories. On this count, new institutionalism is particularly attractive in that it allows multiple perspectives to flourish, thereby encouraging scholars in the increasingly divided discipline of political economy to find commonalities. Of course, many scholars disagree with the basic assumptions of new institutionalism. Some consider it equivalent to rational-choice institutionalism. Rationality implies that human behavior is consequence governed, and that actors estimate their costs-benefits prior to action. In contrast, many scholars

[7] In this context see Weimer (1997) where multiple theories have been employed (economics, public choice, and distributional) to have better specified understanding of the privatization experiences in Eastern Europe and the former Soviet Union.

view human behavior to be rule governed (March and Olsen 1989; Sandholtz 1999). Still, others give primacy to ideas and discourse in shaping human behavior (Hall 1986; Wendt 1992). For them, ideas are not mere intervening variables since preferences are socially constructed.

Leaders matter

New institutionalists assume methodological individualism implying that all actors are ontologically equal; it does not imply that all individuals are alike. Individuals differ in their preferences, their access to resources, their abilities to employ such resources, and their capacities to influence outcomes. Some individuals have greater capacities and willingness to influence institutional evolution in favor of Type 2 or other policies. The discussions on underground storage tanks, the 33/50 program, and Responsible Care suggested that such leaders demonstrate capacities to convince policy skeptics to revise their assessments of the long-term benefits of Type 2 policies.

This study does not view power-based processes reflective of leadership. Structurally advantaged individuals such as hierarchical superiors may force the adoption of Type 2 policies but such interventions do not lead policy skeptics to revise their assessments of Type 2 policies. Even recently I observed continued opposition in Baxter to external audits by Arthur D. Little (ADL). The ADL consultant was more anxious about his presentation to the Divisional Environment Managers (DEMs; policy skeptics) than to the Environmental Review Board which is the highest decision-making body on environmental issues. After attending his presentation to the DEMs, I better appreciated his reasons for being anxious; some DEMs continue to oppose external audits by ADL. Hence, I infer that although external audits have been mandated by senior management, many DEMs have not revised their preferences against them.

Similarly, though a senior manager held up the implementation of the Community Outreach Program under the aegis of Responsible Care in Lilly, policy supporters did not give up their efforts. After about eighteen months, once the policy skeptic retired, policy supporters persuaded Lilly to implement community outreach programs.

A focus on leadership is also helpful in strengthening the dialogue between political economy and organizational theory. Though both political economists and organizational theorists study collective action, their intellectual agendas appear to have drifted apart. Political economists tend to study the dynamics between politics and economics. A study of politics is often (incorrectly) equated to the study of governments, and a study of economics to the study of markets. Consequently, the intellectual arena for many political economists is limited to studying two institution types: markets and governments. Some political economists, especially public choice scholars, tend to employ the tools of economics to study politics (Mueller 1989). On the other hand, organizational theorists tend to employ the tools of sociology. Sociologists also tend to focus less on markets. Due to differences in methodological tools and research focus, the level and quality of dialogue between economists and sociologists leaves much to be desired. This drift is unfortunate since both political economists and organizational theorists study collective action; they have much to contribute to each other. New institutionalism, especially the Ostrom version, combines insights (at least) from political science, economics, and sociology. Therefore, it has the potential to initiate a dialogue between organizational theory and political economy. Since leadership is a highly researched subject in organizational theory, this is one area where such cross pollination can begin.

Does studying leadership imply that our research will become focused on examining preferences and strategies of leaders, and not of "ordinary folks"? If so, will it distort our understanding of social processes? Such debates have reverberated in other contexts as well. For example, historians such as Ambrose (1967) emphasize the role of leaders, while others such as Zinn (1995) and Appleby, Hunt, and Jacob (1994) emphasize the contributions of popular movements.[8] In international relations, realists such as Morgenthau (1978) and Kissinger (1964) emphasize the role of leaders in shaping foreign policy. In their study of the Indian Freedom movement, historians such as Majumdar, Raychaudhuri, and Datta (1958) focus on the contributions of Gandhi, Nehru, Bose, and other stalwarts, while "subaltern historians" emphasize the role of popular movements (Sarkar 1989).

Clearly, for studying social processes at the firm level or country level, one needs to examine the contributions of both leaders and "ordinary folks," as well as the structures within which they function. Political economists focus on structures (with their emphasis on institutions) as well as "ordinary folks" (an implication of methodological individualism). Some scholars also examine the roles of structurally advantaged actors in shaping social processes and I have termed such explanations power-based theories. However, leadership seems to be a relatively neglected area in the study of political economy.

Would introducing leadership as a key variable in political economy lead to elitist explanations of social processes? Such concerns are unwarranted as we are interested in understanding the phenomenon of leader-

[8] I thank Cynthia Yaudes for this point.

ship, how it affects collective action, and the evolution of institutions. There is no assertion that only certain types of people can display leadership. What is important is that leadership may influence the organization and form of collective action, whether in small self-governing communities or in global corporations.

Policy implications

This book examined an important policy issue concerning firms' responses to environmental issues. It is often believed that firms are the main agents of environmental degradation: firms cut corners to save on environmental costs, and, as a result, there is an inherent conflict between environmental sustainability and firms' objectives. This study challenges the simplistic notions that suggest that firms will invariably adopt policies that minimize the costs borne by them.

Two policy implications flow from this book. First, factors external to firms (and these could be influenced by policymakers) may have a critical, though *not a deterministic*, influence on environmental policy processes within firms. Second, firms often adopt those Type 2 policies that require fewer changes in their internal organizations and institutions, especially at the collective-choice level. This has important implications for policy design, again a variable that could be manipulated by policymakers.

The role of external factors

In contrast to the neoclassical economic theory (as well as sociological institutional and stakeholder theories) that places *exclusive* emphasis on external factors, this study emphasizes the importance of *both* internal processes and external factors in influencing whether or not firms adopt Type 2 policies. Clearly, external factors are important since they influence the "rules" and "norms" (Crawford and Ostrom 1995) within which firms conduct their business. They may also directly influence managerial behavior. For example, individual managers in US firms can face civil and criminal prosecution for environmental violations by their firms. However, managers may not always take the external institutions as given. As the literature on regulation points out, sometimes managers may even shape them. Stigler (1971) argues that firms may lobby for regulations in certain circumstances. Bernstein (1955) suggests that business firms may "capture" governmental agencies that are supposed to regulate them and Kolko (1963) provides an interesting account of how the railroad industry "captured" the regulatory agencies.

Previously, environmental policies were classified as compliance driven (Type 3 and Type 4) or as beyond-compliance (Type 1 and Type 2). External institutions (laws and regulations in our case) alone are sufficient to predict firms' responses to compliance driven policies. However, external factors may also influence adoption or non-adoption of Type 2 policies by impacting firms' internal policy processes. In particular, three attributes of the external environment are important: "power," "legitimacy," and "urgency" (Mitchell, Agle, and Wood 1997) in the context of a given policy. External stakeholders are powerful if they can compel managers to do something that the managers would not have done otherwise. Legitimacy implies that managers view the actions or demands of the external stakeholder as appropriate within a given set of norms and beliefs. The urgency dimension suggests that managers believe that there is a need for immediate action in response to the demands or requests of the external stakeholders. Managerial perceptions of the three attributes can be expected to vary within and across policies. With Type 2 policies creating a political space for "discursive struggles," policy supporters and policy skeptics compete to sway managerial perceptions one way or other.

As suggested in table 5.2, if managers perceive external factors which promote a given Type 2 policy as legitimate, urgent, and significant in shaping their firm's market and non-market environment, policy supporters have greater *incentives* to mobilize internal coalitions. Since efforts of policy supporters have greater *credibility* among policy skeptics, there is a greater likelihood that internal coalitions in support of a given Type 2 policy will emerge.

Of the six Baxter's policies the book examined, five were adopted (underground tanks, green products, 33/50, internal, and external audits). In three of the five adopted policies, Baxter's managers perceived external actors encouraging policy adoption as important (underground tanks, green products, and 33/50). For internal and external audits, however, they perceived external factors as important but discouraging policy adoption. Baxter did not adopt ISO 14000, primarily because managers considered external factors encouraging this policy to be relatively unimportant. Also, policy supporters could not capture the top management and have it mandate policy adoption.

Similarly, of the six Lilly's policies the book examined, four were adopted (underground tanks, 33/50, phase two of Responsible Care, and internal audits). In three of these four policies, managers perceived external factors encouraging policy adoption as significant (underground tanks, 33/50 and phase two of Responsible Care). For internal audits, however, managers viewed external factors as important but discouraging

Table 5.2. *External factors*

Policy	Outcome/Process	Importance of external factors
Baxter		
Underground tanks	Adopted: leadership	High; positive
EPA's 33/50	Adopted: leadership	High; positive
Responsible Care	n.a.	n.a.
ISO 14000	Not adopted	Low; positive
Internal audits	Adopted: Power	High; negative
External audits	Adopted: Power	High; negative
Green products	Adopted: leadership	High; positive
Lilly		
Underground Tanks	Adopted: leadership	High; positive
EPA's 33/50	Adopted: leadership	High; positive
Responsible Care	Initially not adopted; Finally adopted: leadership	High; positive
ISO 14000	Not adopted	Low; positive
Internal audits	Adopted: leadership	High; negative
External audits	Not adopted	High; negative
Green products	n.a.	

Summary

Importance of external factors	*Adopted*	*Not adopted*
High and positive	Underground tanks (Baxter, Lilly) 33/50 (Baxter, Lilly) Responsible Care (Lilly) Green products (Baxter)	Responsible Care (Lilly)
High and negative	Internal audits (Baxter, Lilly) External audits (Lilly)	External Audits (Baxter)
Low and positive		ISO 14000 (Baxter, Lilly)

policy adoption. Similar to Baxter, Lilly did not adopt ISO 14000 since managers considered external factors as relatively unimportant. Further, policy supporters could not influence the top management and have it mandate policy adoption.

Both Lilly and Baxter adopted the 33/50 program primarily because its sponsor, the EPA, had the capacity to significantly influence these firms' business environments. The EPA was also perceived as a legitimate actor and the need to subscribe to 33/50 program was crucial to retain public confidence. To recap, 33/50 aimed to reduce emissions of TRI chemicals. The TRI database was accessible to the public, and environmental groups used it to identify key polluters in a given county and state. The EPA's sponsorship of this program facilitated the emergence of internal coalitions (leadership-based policy). Even though 33/50 was a voluntary program, its supporters within Baxter and Lilly argued that its adoption was necessary for their firms to remain in the good books of the EPA. The perception among managers of the coercive character of the EPA created incentives for and imparted credibility to the actions of policy supporters to rally policy skeptics. Policy supporters invoked their firms' experiences with the EPA in the late 1980s and succeeded in convincing the policy skeptics that adopting 33/50 program would create substantial goodwill for their firm within the EPA as well as with other external stakeholders. However, not every US firm perceived the EPA as being sufficiently coercive and/or legitimate, and few believed that adopting 33/50 would keep their firms in the good books of the EPA. As a result, only 13 percent of firms releasing TRI chemicals adopted the 33/50 program.

Leadership-based processes also explain Lilly's and Baxter's adoption of Type 2 policies on replacing underground tanks. Policy supporters within these firms stressed that even a single leak from an underground tank could seriously harm their firm's reputation. Since both Lilly and Baxter are in the health-care business where a firm's reputation is very important, this could inflict substantial long-term (non-quantifiable) financial damage on them. In Lilly, the coercive character of the US legal system, particularly the award of punitive damages, was important in rallying support on this policy. Policy supporters also stressed that Lilly must ensure that communities in its facilities' vicinity view Lilly as a responsible and trustworthy corporate citizen. Nearly 90 percent of Lilly's tanks are located in its Tippecanoe and Clinton facilities in Indiana. Previously, there had been unsubstantiated complaints that leakages from underground tanks contaminated the region's aquifers. Hence, the local communities (external stakeholder) needed reassurance that Lilly's storage tanks did not leak; installing expensive above-the-ground tanks (Type 2 policy) enabled visitors to Lilly's facilities to visually inspect the storage tanks.

In part, Lilly's enthusiastic support for Responsible Care can be attributed to its desire to be a major player in the Chemical Manufacturers Association (CMA). Lilly is a show-case example of the successful implementation of Responsible Care. Since the CMA has an important voice on policies impacting the chemical industry, the major players within the CMA have opportunities to pursue their firms' agenda under its aegis.

Policy supporters highlighted the significant payoffs of adopting Responsible Care in terms of increased credibility with CMA member firms, thereby acquiring greater ability to influence CMA's agenda.

In the "green products" case (chapter 4), the book argued that the nature of the business portfolio (external factor) critically influences policies on the manufacturing and marketing of green products. Consequent to the EPA's proposal on air-quality standards and the rising costs of disposing wastes, Baxter's customers began talking about packaging reductions and reducing PVCs. This created an opportunity for supporters of green products to argue that Baxter has important business reasons to be proactive and to consider marketing products that offer environmental benefits as a distinctive product offering. In contrast, since Lilly manufactures and markets ethical drugs, its customers (doctors) potentially have little interest in the greenness of drugs; they look for attributes such as efficacy, cost of treatment, and availability. Hence, for doctors, the greenness of Lilly's products and corporate policies are "hygiene" factors (their absence may dissuade them from prescribing Lilly's products), but not "motivators" (their presence may not persuade them to prescribe Lilly's products).

This discussion has important implications for the environmental policy discourse since "greening" can occur at the level of the firm as well as the level of the product; and one may not always lead to the other. An important question therefore is: what kind of greening should be encouraged? For example, the General Agreement on Tariffs and Trade (GATT), now the World Trade Organization (WTO), allows countries to subject its imports to product standards but not process standards (Esty 1994). Thus, GATT creates incentives for firms to "green" their products but not their management systems. On the other hand, if ISO 14000 certification becomes a *de facto* requirement for exporting to the European Union, firms will have greater incentives to green their management systems.

Policy supporters face a difficult task if policy skeptics and the top management do not consider external factors to be important, legitimate, and urgent; particularly in terms of the external stakeholders' abilities to impose significant excludable costs or to offer significant excludable benefits. It impedes their efforts to mobilize coalitions or to "capture" the top management. Lilly's and Baxter's wait-and-see policies on ISO 14000 (at least, until mid 1996) were partially due to the non-coercive character of the International Organization for Standardization, the sponsor of ISO 14000 (the internal factor impeding the adoption of ISO 14000 being the emergence of a "green-wall"). The policy implication is that ISO 14000 must provide excludable benefits or impose excludable costs on firms and their managers. Only then can leadership-based or power-based processes be expected to empower policy supporters. For example, if the EPA lends visible support to ISO 14000 standards, perhaps US firms may consider ISO 14000 more seriously. The EPA's stand against granting attorney–client privilege to environmental audits (and ISO 14000 involves such audits), makes ISO 14000 even less attractive to US firms.

The discussion on environmental audits offered yet another perspective on the influence of external factors on intra-firm dynamics. As discussed in the case on environmental audits (chapter 4), many states have passed laws that encourage firms to undertake self audits. The EPA ostensibly supports self audits. However, it vehemently opposes granting attorney–client privilege to information gathered during such audits, thereby discouraging self audits. Given the coercive powers of the EPA, external factors appear to discourage firms from undertaking audits.

Both Lilly and Baxter have established internal audit programs. How does one explain this anomaly? Policy supporters successfully portrayed audits as tools to facilitate compliance and to improve environmental management systems. They contended that since their firms were committed to comply with or exceed the requirements of all applicable laws and regulations, the firm had nothing to fear from audits. And if there were instances of unintentional violations (the assumption being that there are not intentional systemic violations), firms would rather report them themselves, and correct them as soon as possible. Policy supporters perceived that, in the long-term, the EPA and other external actors that opposed such audits would realize that firms are really not misusing them, and will therefore begin to view audits more favorably. As a result, proactively establishing internal audit systems will have long-term payoffs.

Demythologizing beyond-compliance

An important finding is that if policy skeptics and top management perceive that external organizations encouraging a Type 2 policy can impose excludable costs or provide excludable benefits, they are more amenable to listening to policy supporters. The case studies described how the managerial perceptions of the EPA's coercive character enabled policy supporters to successfully argue for Type 2 policies. Adopting Type 2 policies appears to have been encouraged by existing regulatory regimes. Existing compliance standards set the base level upon which beyond-compliance initiatives take place.

Arguably, firms adopt beyond-compliance initiatives primarily to preempt even more stringent regulations or to shape future regulations.

To some extent this is true since many policy supporters argued that firms need to adopt Type 2 policies if they wish remain a credible player in the environmental policy discourse. However, one needs to be careful in attributing adoption of *all* Type 2 policies to such motivations. Firms face collective action dilemmas in adopting Type 2 policies; only concerted action by firms can potentially ward off tougher regulations or shape future regulations. The preemption argument may hold for industry-level initiatives such as Responsible Care and ISO 14000 that explicitly seek to preempt stringent regulations. By converting goodwill for firms among regulators and other external actors from a public good to a club good, industry-level initiatives mitigate collective action dilemmas (chapter 2). Consequently, firms have greater incentives to adopt such Type 2 policies. It is, however, difficult to explain how Lilly's adoption of the most expensive route to replace underground tanks could ward off stringent regulations or how Baxter's use of external auditors could convince the EPA to adopt a more industry-friendly approach by granting attorney–client privileges on environmental audits.

Organizational change

Type 2 policies requiring significant organizational change are less likely to be adopted by leadership-based processes. By creating new institutions and structures, or by reallocating responsibilities within existing structures, organizational changes upset the status-quo. Actors "losing" from such changes have incentives to oppose them. If the organizational changes to implement such a policy are "significant" (see below), it becomes difficult to pacify the losers without diluting the integrity of the proposed policy. As summarized in table 5.3 below, policies requiring significant changes may not be adopted by leadership-based processes since it is difficult to induce losers to cooperate. If they are adopted at all, they reflect power-based processes whereby structurally powerful policy supporters were able to impose their will on policy skeptics. Though policy skeptics remain dissatisfied, they are forced to accept the new policy.

Of the six Baxter's policies the book examined, organizational changes were significant only for environmental audits – both internal and external. Not surprisingly, Type 2 policies on internal and external audits were adopted by power-based processes. Organizational changes were less significant for the other four, three of which were adopted by leadership-based processes (underground tanks, 33/50, and green products), and one which was not adopted (ISO 14000).

Of the six Lilly's policies examined, organizational changes were

Table 5.3. *Organizational change*

	Outcome/Process	Organizational change
Baxter		
Underground tanks	Adopted: leadership	Low
EPA's 33/50	Adopted: leadership	Low
Responsible Care	n.a.	n.a.
ISO 14000	Not adopted	Low
Internal audits	Adopted: Power	High
External audits	Adopted: Power	High
Green products	Adopted: leadership	Low
Lilly		
Underground tanks	Adopted: leadership	Low
EPA's 33/50	Adopted: leadership	Low
Responsible Care	Initially not adopted	High
	Finally adopted: leadership	Low
ISO 14000	Not adopted	Low
Internal audits	Adopted: leadership	Low
External audits	Not adopted	High
Green products	n.a.	n.a.

Summary

Outcome/Process	Organizational changes are significant	Organizational changes are not significant
Policy is not adopted	External audits (Lilly) Responsible Care (Lilly)	ISO 14000 (Baxter, Lilly)
Policy is adopted by power-based processes	Internal and external audits (Baxter)	
Policy is adopted by leadership-based processes	Internal audits (Lilly) Responsible Care (Lilly)	Underground tanks (Baxter, Lilly) 33/50 (Baxter, Lilly) Green products (Baxter)

significant in two of them (external audits and phase one of Responsible Care). Consequently, these policies were not adopted. Organizational changes were less significant for the other four policies, three of which were adopted by leadership-based processes (underground Tanks, 33/50, and the second phase of Responsible Care), and one which was not adopted (ISO 14000).

Whether or not institutional changes are significant requires examining the nature of the institutions. The Institutional Analysis and Development (IAD) framework identifies three nested levels of institutional analysis: operational choice, collective choice, and constitutional choice (Kiser and Ostrom 1982). Constitutional choice level changes are the most difficult to implement while operational choice level changes are the easiest. The former alter the "status quo" most significantly since all future collective choice decisions would be affected by such changes. Consequently, they create incentives for "losers" to tenaciously resist them. In the early 1990s, Baxter established environmental auditing systems and invited Arthur D. Little (ADL) to define state-of-the-art standards. This action deprived Baxter's facility managers of a crucial voice in defining environmental standards. As a result, they opposed the concept of state-of-the-art standards and the invitation to ADL to define them. Such collective choice level changes are adopted only if their supporters are hierarchical superiors having the abilities to ignore opposition. Baxter's policy had vocal and decisive support from Vernon Loucks, Baxter's Chief Executive Officer, and Marshall Abbey, Chairman of Baxter's Environmental Review Board who virtually mandated the adoption of this policy.

In contrast, Lilly's policy on internal audits did not ignite internal opposition since it did not upset the status quo. Unlike Baxter, Lilly defined its auditing standards internally with contributions from both corporate and facility managers. In the early 1990s, Lilly employed environmental audits primarily as a compliance tool, and only in recent years have they been employed to evaluate management systems as well. Such changes in management systems create significant conflicts only if they alter organizational structures or reallocate functional responsibilities. In contrast to Lilly, Baxter did not adopt such an incremental policy on audits; rather, the compliance and management systems aspects were emphasized from the very beginning.

Operational choice level decisions do not significantly impact inter-manager interactions since they do not initiate new policies and they only marginally reallocate responsibilities among existing managers. In effect, they do not create significant losers. This is an important reason why the underground storage tank program and the 33/50 program in Baxter and Lilly, and the marketing of green products in Baxter, met with little resolute opposition. These programs did not require creating new permanent structures (although implementation committees were formed in both these firms) and therefore did not redistribute responsibilities or budgets within these firms. With managers not feeling professionally threatened, they had few incentives to tenaciously oppose the new policies.

On the other hand, some of Lilly's managers initially opposed Responsible Care, especially its Code on Community Awareness and Emergency Response. They questioned its usefulness and legitimacy. This Code required CMA members to develop Community Outreach Programs. The policy skeptics argued that Lilly had little reason to share information on its manufacturing operation since it adhered to all applicable laws and regulations. Further, communities in the vicinity of Lilly's facilities had little knowledge with which to appreciate the technicalities of manufacturing processes. A sharing of technical information with such communities may lead them to make unwarranted conclusions about the safety aspects and environmental impacts of facilities' operations. Opposition to Responsible Care stemmed from the requirement for Lilly to change its policy (collective choice level institution) on outsiders' access to information on Lilly's internal operations. Responsible Care required significant changes in a major policy and the affected internal managers had little voice in shaping this change. Policy skeptics viewed the code as fundamentally redefining the relationship between their firm and communities living in the vicinity of its facilities, and felt threatened by this redefinition.

This discussion has important implications for policy design: policymakers should consider the impact of a proposed policy on organizational politics and the incentives for managers to support or oppose it. Policies that seek to draw upon firms' extant structures and institutions have greater likelihood of adoption and successful implementation. In contrast, if Type 2 policies require significant changes at the constitutional choice and collective choice levels, they will probably encounter stiff internal opposition. And perhaps, such opposition may only be overcome by employing power-based processes.

Limitations and future research

This study employed a new-institutionalist perspective to examine environmental policymaking in Baxter and Lilly. It integrated insights from sociological institutional theory and stakeholder theory with power-based and leadership-based theories. Intra-firm policy processes and inter-manager interactions can also be examined by employing other theories and frameworks such as resource-based (Penrose 1959; Peteraf 1993), evolutionary (Nelson and Winter 1982), garbage can (Cohen, March, and Olsen 1972), and ecology (Hannan and Freeman 1977). These theories do not employ the same set of assumptions as new-institutionalism. In particular, since most of them do not subscribe to methodological individualism, they do not treat institutions (policies of firms) as conscious artifacts that can be traced to the preferences, endowments, and strategies of individual actors. These theories could therefore suggest different explanations for why firms selectively adopt Type 2 policies. Hence, this study should be viewed as a modest attempt to examine the phenomenon of beyond-compliance policies by employing one of many possible perspectives.

I focused on Type 2 policies since they appear anomalous to neoclassical economic theory. My conclusions are not generalizable to all beyond-compliance policies that include Type 1 as well. On this count, this book should not be interpreted as a general critique of the neoclassical theory.

Transaction cost theories are an integral part of the new institutionalist framework. Coase (1937) and Williamson (1975, 1985) argue that firms and other institutions arise to minimize transaction costs associated with market exchanges. Transaction cost theory is therefore employed to understand structural changes such as evolution of "hierarchies." In general, many argue that institutions are efficiency-enhancing artifacts and they evolve since actors wish to reduce the transaction costs associated with collective action (for a review, see Eggertsson 1990). Future research could also focus on applying transaction cost theories to understand evolution of micro-firm-specific policies such as Type 2 policies. For example, it could be hypothesized that firms adopt those Type 2 policies that reduce the sum of transaction cost and transformation cost. Since the standard capital budgeting techniques cannot capture the impact of policies on transaction costs in the future, some ways to measure transaction costs are required to test such hypothesis.

Previously, I discussed the need to demythologize the adoption of beyond-compliance policies which should be understood only in the context of extant laws and regulations and the incentives for preempting more stringent laws or shaping future laws. Adopting Type 2 policies is a self-interested response by managers and their firms. Beyond-compliance policies are adopted in other areas as well such as workplace safety and consumer safety. It would therefore be instructive to compare the processes of beyond-compliance policymaking across issue areas within and across firms and examine the validity of my analysis. Further, I have not "unpacked" regulatory agencies (such as the EPA) and implicitly treated them as monoliths. An area of further research could be to examine how differences in regulatory styles within a given governmental agency impact dynamics within and across firms on beyond-compliance policies.

This study focused on policy adoption only. It has not examined the durability and efficacy of power-based policies versus leadership-based policies. Arguably, power-based policies are effective in the short-run but not in the long-run. Only a consensual route of policy adoption succeeds in the long-run given that skeptics often have ways to oppose, and if driven to desperation, even to sabotage it. On the other hand, it could be argued that if top management's intent is clearly communicated, and power-based policies are able to meet their stated objective, the skeptics may become reconciled to new rules of the game. There are theoretical reasons to argue in favor as well as against power-based policies in the context of their long-run durability and efficacy. However, this is an important issue worthy of more research.

In examining inter-manager interactions, the book classified managers as policy supporters and policy skeptics. Somewhat in the tradition of Samuelson's (1947) 'revealed preference,' it inferred managerial preferences for Type 2 policies from their behaviors. Methodologically, however, this is an imperfect way of assessing preferences since preferences and exhibited behaviors may not have direct correspondence. In other words, the same set of preferences can translate into different behaviors given varying incentives. In some instances, actors may not have opportunities to express their preferences or may choose not to do so due to fears of retribution.

Being conscious of such pitfalls, I examined the reasons why managers support or oppose a given policy. During my interviews, I gathered information on the professional backgrounds of managers working on environmental issues across functional areas. From these interviews it was clear that most of them had strong personal commitments to environmental issues. This was not surprising since they chose to work on environmental issues; not because of organizational-level incentives such as special allowances or faster promotions. Many of these managers also claimed to be "environmentalists": some of them claimed to support the Sierra Club and other environmental groups financially. I also observed posters on environmental issues adorning offices of most of these managers. Most of them are very active in Earth Day celebrations. Such a display of support for environmental issues and my numerous discussions with them leads me to infer that most of these managers indeed hold strong beliefs on environmental issues. Thus, I expect that such managers would support Type 2 policies. On the other hand, I also expect "losers" from any organizational change that results from a Type 2 policy to assume the roles of skeptics. These losers could belong to environmental departments or work on environmental issues in other departments. Nevertheless, the strategy of inferring preferences from behaviors is not methodologically perfect.

This book argued that Type 2 policies are adopted only through power-based or leadership-based processes. It could be argued that leadership-based processes are in fact power-based processes where power is exercised subtly in terms of shaping opinion. As discussed in chapter 2, scholars following the perspectives of Gramsci (1988) and Foucault (1970) could also be expected to argue along the same lines. For example, it could be argued that efforts of policy supporters such as having their firm's Washington lobbyists make presentations on the desirability of adopting Type 2 policies (as in underground tanks) are really strategies to manipulate opinions of the skeptics. In this context, power is being subtly exercised.

If power is defined in such an all-encompassing manner, it becomes impossible to make falsifiable predictions about whether or not power has been exercised. The book defined the outcomes of power-based processes in a specific way: preferences and exhibited behaviors of policy skeptics do not change. If the ability to influence preferences (as in leadership-based processes) represents an exercise of power, then leadership is indeed a manifestation of power. I, however, have chosen not to define power in this manner.

Power-based and leadership-based processes are indeed different for the following reasons. First, in power-based processes the evidence of dissent is very clear. Such continued opposition of policy skeptics became even more evident during interviews. Once I had won their confidence, especially in terms of protecting confidentiality, I came across a wealth of information on their continued negative perceptions about Type 2 policies. It then became easier for me to separate the instances where preferences had changed due to induced cooperation from where preferences had not changed at all. Further, in both these firms, most managers are fairly independent and outspoken; if they disagree, they speak out. Such public displays of disagreements have become even more pronounced since both of these firms are attempting to decentralize and to empower their division- and facility-level managers.

The book suggests that power-based and leadership-based explanations *together* explain why firms selectively adopt Type 2 policies. It could, however, be argued that senior managers first employ a leadership-route, and if it does not work out, resort to power-based processes. Hence, power-based and leadership-based processes are sequential. I disagree with this as well. When policy supporters in Baxter decided to invite external auditors to define and then to audit facilities against the state-of-the-art standards, they adopted a power-based route: Vernon Loucks and Marshall Abbey mandated its implementation. Why was a leadership-based route not adopted? The reason is that since the policy supporters had committed a very tight implementation schedule to the Environmental Review Board, they did not want to experiment with a potentially time-consuming consensual route. As a result, they opted for employing the power of top management fiat. This suggests that the strategy employed by policy supporters is not sequential; it depends on a variety of factors such as the time table for implementation, the level of perceived opposition, and the perceived success of employing fiat.

This book operationalized procedural efficiency in terms of the most well-accepted and widely used procedure of investment analysis – capital budgeting – employed by large US firms. To make *ex ante* falsifiable predictions that firms are efficient institutions, efficiency needs to be defined in a precise manner. Otherwise, most policies would be classified *ex post* as being efficient, robbing the concept of efficiency of its analytical power. Other investment procedures such as full-cost accounting and life-cycle analysis lack acceptability among managers, were not employed by either Lilly or Baxter, and, importantly, do not lead to the maximization of shareholders' wealth, the avowed objective in the neoclassical theory. The book, therefore, operationalized procedural efficiency in terms of capital budgeting. This procedure seeks to maximize shareholder's wealth, an important objective geared for minimizing principal–agent conflicts in large corporations where principals (shareholders) do not have a direct say in the running of the firm and are vulnerable to agent (managers) abuse. However, in owner-managed businesses where such principal-agent conflicts are not significant, capital budgeting may not be employed. It is also difficult to identify other procedures that are consistently employed by owner-managers to *ex ante* estimate profitability of projects. Nevertheless, this does not imply that such businesses are inefficient. Thus, the findings and conclusions of this book are applicable only to large corporations that employ capital budgeting as an established tool for investment analysis.

In spite of their widespread acceptability in large US corporations, discounted cash flow techniques have been widely criticized. Some scholars argue that an over-reliance on them has lead to decline of America's competitive position (Donaldson 1972; Hayes and Abernathy 1980; Hout, Porter, and Rudden 1982). They point out methodological biases in capital budgeting that distort investment decisions, thereby forcing managers to adopt short-term horizons (Hodder and Riggs 1985; Kulatilaka and Marcus 1992). Clearly, with different national systems of industrial organization, capital budgeting is perhaps not a widely accepted tool in say the Japanese *Keiretsus* (Hodder 1984). Most Japanese firms are overcapitalized by US standards, an outcome that would be discouraged by capital budgeting procedures. It is also suggested that Japanese managers have long time horizons, while capital budgeting encourages shorter time horizons. Further, since Japanese managers are perceived as being relatively insulated from the pressures of the stock markets, it is suggested that they would not be unduly worried about ensuring adequate rates of return on investment. Kester notes that:

> Japanese corporate governance emphasizes the reduction in transaction costs associated with self-interested opportunism and investment in relations-specific asset. This strategy fosters the building of stable, long-term commercial relationships among transacting companies, *although general (i.e., noncorporate, nonlending) shareholders are often forced to bear potentially substantial agency costs.* Anglo-American corporate governance, in contrast, emphasizes the reduction in agency costs associated with the separation of ownership from control, relying more heavily on formal, legalistic mechanisms to order commercial relationships among transacting parties (1996: 108; italics mine).

Hence, capital budgeting may be inadequate, if not the wrong route, to operationalize efficiency in the Japanese or other non-US systems of industrial organizations. On this count, my findings and conclusions may lack generalizability to non-US contexts.

As discussed in chapter 1, the 1993 UNCTAD benchmark survey suggests that environmental programs of MNEs are critically influenced by three factors: sales volume, nationality of their parent firm, and industry type. Baxter and Lilly share these characteristics in that both have 1996 sales greater than $4.9 billion, both have American parents, and both are in similar industries. On this count, the implication of this research for theory development and policymaking may not be generalizable to firms that vary on one or more of these three attributes. Consequently, future research could investigate the implications of this research by studying firms that vary on these three attributes.

References

Ackerman, R. W. 1975. *The Social Challenge to Business*. Cambridge, MA: Harvard University Press.

Alchian, A. A. 1950. Uncertainty, evolution, and economic theory. *Journal of Political Economy*, 58: 211–21.

Allegiance. 1997. http://www.allegiance.net/COMPNEWS/Feb0397.HTM; page 2 of 3; 04/27/97.

Allison, G. T. 1971. *Essence of Decision: Explaining the Cuban Missile Crisis*. Boston, MA: Little Brown.

Ambrose, S. E. 1967. *Eisenhower and Berlin*. New York: W.W. Norton.

Appleby, J. O., Hunt, L., and Jacob, M. 1994. *Telling the Truth About History*. New York: Norton.

Arkansas Gazette. 1992. March 19: Arkansas Page.

Arora, S. and Cason, T. N. 1996. Why do firms volunteer to exceed environmental regulations? Understanding participation in EPA's 33/50 program. *Land Economics*, 72: 413–32.

Arrow, K. J. 1987. Reflections on the essays. In G. Feiwel (ed.), *Arrow and the Foundations of the Theory of Economic Policy*. New York: New York University Press, pp. 727–34.

Arthur D. Little, Inc. 1995. Hitting the green wall. *Perspectives*.

Atkinson, S. and Tietenberg, T. H. 1991. Market failures in incentive-based regulations: The case of emission trading. *Journal of Environmental Economics*, 21: 17–31.

Barnard, C. I. 1938. *The Functions of an Executive*. Cambridge, MA: Harvard University Press.

Barrett, S. 1991. Environmental regulations for competitive advantage. *Business Strategy Review*, 2: 1–15.

Barton, S. L., Hill, N. C., and Sundaram, S. 1989. An empirical test of stakeholder theory predictions of capital structure. *Financial Management*, 18: 36–44.

Bass, B. M. 1985. *Leadership and Performance Beyond Expectations*. New York: Free Press.

Bates, R. H. 1983. *Essays on the Political Economy of Rural Africa*. Cambridge University Press.

Baumol, W. J. 1959. *Business Behavior, Value, and Growth*. New York: MacMillan.
  1982. *Economic Theory and Operations Analysis*. New Delhi: Prentice Hall.

Baumol, W. J. and Oates, W. 1988. *The Theory of Environmental Policy*. Cambridge University Press.

Baxter Environmental Manual/Baxter. 1994. Deerfield, IL: Baxter Healthcare Corporation

Baxter Healthcare Corporation/Baxter. 1985a. *Travenol Environmental Newsletter.* October. Deerfield, IL: Baxter Healthcare Corporation

1985b. *Travenol Environmental Newsletter.* December. Deerfield, IL: Baxter Healthcare Corporation

1986 *Travenol Environmental Newsletter.* October. Deerfield, IL: Baxter Healthcare Corporation

1988. *Baxter Environmental Review.* September. Deerfield, IL: Baxter Healthcare Corporation

1989. *Baxter Environmental Review.* October. Deerfield, IL: Baxter Healthcare Corporation

1990a. *Baxter Environmental Review.* May. Deerfield, IL: Baxter Healthcare Corporation

1990b. *Environmental Leadership: Remarks by Vernon R. Loucks Jr., Chairman and Chief Executive Officer, Baxter International.* Deerfield, IL: Baxter International, Inc.

1991a. *Baxter Environmental Review.* May. Deerfield, IL: Baxter Healthcare Corporation

1991b. *Baxter Environmental Review.* September. Deerfield, IL: Baxter Healthcare Corporation

1992a. *Serving the Environment: A Call to Action for the Health Care Industry.* Deerfield, IL: Baxter International, Inc.

1992b. *Baxter Environmental Review.* May. Deerfield, IL: Baxter Healthcare Corporation

1992c. *Baxter Environmental Review.* August. Deerfield, IL: Baxter Healthcare Corporation

1993. *Environmental Highlights.* Deerfield, IL: Baxter Healthcare Corporation

1994a. *Annual Report.* Deerfield, IL: Baxter International, Inc.

1994b. *Environmental Performance Report.* Deerfield, IL: Baxter International, Inc.

1994c. *Baxter Environmental,* 2(3). Deerfield, IL: Baxter International, Inc.

1994d. *Baxter Environmental Initiatives for Customers.* Deerfield, IL: Baxter International, Inc.

1995a. *State-of-the-art Environmental Program Report 1994.* Deerfield, IL: Baxter International, Inc.

1995b. *Baxter Environmental,* 3(2). Deerfield, IL: Baxter International, Inc.

1996a. *Environmental Performance Report 1995.* Deerfield, IL: Baxter International, Inc.

1996b. *Internal Memo on "Customer Environmental Initiatives."* May 16. Deerfield, IL: Baxter International, Inc.

1996c. *Annual Report.* Deerfield, IL: Baxter International, Inc.

1997a. http://www.baxter.com/www/financia...port/1996/financialhighlights.html; page 1 of 2; 04/27/97.

1997b. *Environmental, Health and Safety Performance Report.* Deerfield, IL, Baxter International, Inc.

Berle, A.A. and Means, G.C. 1932. *The Modern Corporations and Private Property.* New York: MacMillan.

Bernstein, M. H. 1955. *Regulating Business by Independent Commission.* Princeton, NJ: Princeton University Press.

Bessen, S. M. and Saloner, G. 1988. *Compatibility Standards and the Market for Telecommunication Services.* Santa Monica, CA: Rand.

Boulding, K. E. 1963. Towards a pure theory of threat systems. *American Economic Review,* 53: 424–34.

Bowles, S. and Gintis, S. 1993. The revenge of homo economicus: Contested exchange and the revival of political economy. *Journal of Economic Perspectives,* 7: 83–102.

Buchanan, J.M. 1962. Marginal notes on reading political philosophy. In J. M. Buchanan and G. Tullock. *A Calculus of Consent.* Ann Arbor, MI: University of Michigan Press, pp.307–22.

1965. An economic theory of clubs. *Economica,* 32: 1–14.

Bunge, J., Cohen-Rosenthal, E., and Ruiz-Quintanilla, A. 1995. Employee participation in pollution reduction. New York State School of Industrial and Labor Relations and Cornell University, Mimeo.

Carroll, A. B. 1995. Stakeholder thinking in three models of management morality: A perspective with strategic implications. Reprinted in M.B.E. Clarkson, (ed.), *The Corporation and its Stakeholders.* Toronto: University of Toronto Press, pp. 139–70.

Carson, R, 1962, *Silent Spring.* Greenwich, CO: Fawcett.

Cascio, J. 1994. International environmental management standards – ISO 9000s less tractable siblings. *ASTM Standardization News,* April: 44–9.

Chandler, A. D. 1962. *Strategy and Structure.* Cambridge, MA: MIT Press.

1977. *The Visible Hand.* Cambridge, MA: Harvard University Press.

1980. Government versus business: An American phenomenon. In J. T. Dunlop (ed.) *Business and Public Policy.* Cambridge, MA: Harvard University Press, pp. 1–11.

Charter, M. (ed.). 1992. *Greener Marketing.* Sheffield: Greenleaf Publishing.

Chatterjee, P. and Finger, M. 1994. *The Earth Brokers: Power, Politics, and Development.* London and New York: Routledge.

Chemers, M.M. and Ayman, R. (eds.) 1993. *Leadership Theory and Research: Perspectives and Directions.* San Diego, CA: Academy Press.

Chemical and Engineering News. 1992. Chemical makers pin hope on Responsible Care to improve image. October 5: 13–39.

1993a. Responsible Care program poses challenges for smaller firms. August 9: 9–14.

1993b. New initiatives take aim at safety performance of chemical industry. November 29: 12–36.

Chemical Manufacturers Association/CMA. 1996a. *About CMA.* http://www.cmahq.com/abtcma.html, 12/27/96.

1996b. *News and Issues.* http://www.cmahq.com/news1.html, 12/27/96.

1998. *Management System Verification.* http://www.cmahq.com/cmapgrams; 1/16/98.

1999. *Community Right to Know.* http://www.cmahq.com/issueadvocay.nsf; 03/12/99.

Chemical Week. 1995. Managers gear up for global standards. October 11: 65–6.

Child, J. 1972. Organizational structure, environment, and performance – The role of strategic choice. *Sociology*, 6: 1–22.

Choucri, N. (ed.) 1993. *Global Accord: Environmental Challenges and International Response*. Cambridge, MA: MIT Press.

Clarkson, M.B.E. 1995. A stakeholder framework for analyzing and evaluating corporate social performance. *Academy of Management Review*, 20: 92–117.

Coase, R. H. 1937. The nature of the firm. *Economica*, 4: 386–405.

1988. *The Firm, the Market, and the Law*. Chicago, IL: University of Chicago Press.

1993. The institutional structure of production: 1991 Noble Prize lecture. In O. E. Williamson and S. Winter (eds.), *The Nature of the Firm*. New York: Oxford University Press, pp. 227–35.

Cochran, P.L. and Wood, R. A. 1984. Corporate social responsibility and financial performance. *Academy of Management Journal*, 27: 42–56.

Coddington, W. 1993. *Environmental Marketing: Positive Strategies for Reaching the Green Consumer*. McGraw-Hill.

Cohen, M.D., March, J.G., and Olsen, J.P. 1972. A garbage can model of organizational choice. *Administrative Science Quarterly*, 17: 1–25.

Colborn, T. 1996. *Our Stolen Future: Are We Threatening Our Fertility, Intelligence, and Survival: A Scientific Detective Story*. New York: Dutton.

Commons, J. R. 1968. *Legal Foundations of Capitalism*. Madison, WI: University of Wisconsin Press.

Cornes, R. and Sandler, T. 1996. *The Theory of Externalities, Public Goods, and Club Goods*. Cambridge University Press.

Crawford, S. E. S. and Ostrom, E. 1995. A grammar of institutions. *American Political Science Review*, 89: 582–600.

Cross, J.G. and Guyer, M. J. 1980. *Social Traps*. Ann Arbor, MI: University of Michigan Press.

Currie, R. M. 1995. 33/50 program progress update, December 1995. A note prepared for submission to the EPA, December 20, 1995. Deerfield, IL: Baxter Healthcare Corporation.

Cyert, R. M. and March, J. G. 1963. *A Behavioral Theory of Firm*. Englewood Cliffs, NJ: Prentice Hall.

Dahl, R. A. 1957. The concept of power. *Behavioral Science*, 2: 201–15.

Davis, T. R. V. and Luthans, F. 1980. A social learning approach to organization behavior. *Academy of Management Review*, April: 281–90.

Deming, W. E. 1982. *Quality, Productivity, and Competitive Position*. Cambridge, MA: Massachusetts Institute of Technology, Center for Advanced Engineering Study.

1992. *The World of W. Edward Deming*. Compiled by C. S. Kilian. Knoxville, TN: SPS Press.

Demsetz, H. and Lehn, K. 1985. The structure of corporate ownership: Causes and consequences. *Journal of Political Economy*, 93: 1155–77.

DiMaggio, P. and Powell, W. W. (eds.) 1983. *The New Institutionalism in Organizational Analysis*. Chicago, IL: University of Chicago Press.

Dodd, L. and Schot, R. 1979. *Congress and the Administrative State*. New York: Wiley.

Donaldson, G. 1972. Strategic hurdle rates for capital investment. *Harvard Business Review,* 50: 50–5.
Donaldson, T. and Preston, L.E. 1995. The stakeholder theory of the corporation: Concepts, evidence, and implications. *Academy of Management Review,* 20: 65–91.
Dow, G. 1987. The function of authority in transaction costs economics. *Journal of Economic Behavior and Organization,* 8: 13–38.
Dugger, W. 1983. The transaction cost analysis of Oliver Williamson: A new synthesis? *Journal of Economic Issues,* 17: 95–114.
Dunning, J. 1993. *The Globalization of Business.* London: Routledge.
Edwards, R. 1979. *Contested Terrain: The Transformation of the Workplace in the Twentieth Century.* New York: Basic Books.
Eggertsson, T. 1990. *Economic Behavior and Institutions.* Cambridge University Press.
1996. The old theory of economic policy and New Institutionalism. Presented at the Workshop on Economic Transformation, Institutional Change, Property Rights, and Corruption, March 7–8, 1996, Washington, DC.
Environmental Protection Agency/EPA. 1994. *Questions and answers about the 33/50 program.* http://es.inel.gov/partners/3350/q-a.html, 12/11/96.
Enzi, M. 1997. EPA's obstruction of pollution control. *Wall Street Journal,* April 13: A 18.
Esty, D. C. 1994. *Greening the GATT.* Washington, DC: Institute for International Economics.
Etzioni, A. 1988. *The Moral Dimension.* New York: Basic Books.
Fama, E. 1980. Agency problem and the theory of the firm. *Journal of Political Economy,* 88: 288–307.
Fiedler, F. E. 1967. *A Theory of Leadership Effectiveness.* New York: McGraw Hill.
Fischer, K. and Schot, J. (eds.) 1992. *Environmental Strategies for Industries.* Washington, DC : Island Press.
Follett, M. P. 1940. Constructive conflict. In H.C. Metcalf and L. Urwick (eds.), *Dynamic Administration.* New York: Harper and Row Publishers, pp. 30–49.
Foucault, M. 1970. *The Order of Things.* Translated from French. London: Tavistock Publishers.
Freeman, R.E. 1984. *Strategic Management: A Stakeholder Approach.* Boston: Pittman.
Fri, R. W. 1992. The corporation as a nongovernmental organization. *The Columbia Journal of World Business,* 27: 91–5.
Friedman, M. 1970. The social responsibility of business is to increase its profits. *New York Times Magazine,* September 13: 32–3, 122–6.
Furubotn, E. G. and Richter, R. 1991. The new institutional economics: An assessment. In E. G. Furubotn and R. Richter (eds.), *The New Institutional Economics.* College Station: Texas AandM University Press, pp. 1–32.
Gabel, H. L. 1994. From market failure to organization failure. *Business Strategy and the Environment,* 3.
Georgopolous, B. S., Mahoney, G.M. and Jones, N. W. 1957. A path-goal approach to productivity. *Journal of Applied Psychology,* December: 345–55.
Ghoshal, S. and Moran, P. 1996. Bad for practice: A critique of the transaction cost theory. *The Academy of Management Review,* 21: 13–47.

Graen, G., Novak M., and Sommerkamp, P. 1982. The effects of leader-member exchange and job design and productivity and satisfaction: Testing a dual attachment model. *Organizational Behavior and Human Performance,* 30: 10–131.

Gramsci, A. 1988. *An Antonio Gramsci Reader.* D. Forgacs (ed.) New York: Schocker Books.

Granovetter, M. 1985. Economic action and social structure: The problem of embeddedness. *American Journal of Sociology,* 91: 481–510.

Greene, C. N. 1975. The reciprocal nature of influence between leader and subordinate. *Journal of Applied Psychology,* 60: 187–93.

Griffin, J. J. and Mahon, J.F. 1997. The corporate social performance and corporate financial performance debate: Twenty-five years of incomparable research. *Business and Society,* 36: 5–31.

Hahn, R. and Nell, R. 1982. Designing a market for tradable emission permits. In W. Megat (ed.), *Reform of Environmental Regulations.* Cambridge, MA: Ballinger.

Hajer, M.A. 1995. *The Politics of Environmental Discourse.* Oxford: Clarendon Press.

Hall, P. A. (ed.) 1986. *The Political Power of Economic Ideas.* Princeton, NJ: Princeton University Press.

Hannan, M. T. and Freeman, J.H. 1977. The population ecology of organization. *American Journal of Sociology,* 82: 929–64.

Hardin, G. 1968. The tragedy of commons. *Science,* 162: 1243–48.

Hart, S. 1995. A natural resource-based view of the firm. *Academy of Management Review,* 20: 986–1014.

Hayek, F. A. 1945. The use of knowledge in society. *American Economic Review,* 35: 519–39.

1955. *The Counterrevolution of Science.* New York: Free Press.

Hayes, R. and Abernathy, W. 1980. Managing our way to economic decline. *Harvard Business Review,* 58: 67–77.

Herzberg, F. 1966. *Work and the Nature of Man.* Cleveland: World Publishing Company.

Hirshleifer, J. 1988. *Price Theory and Its Applications.* 4th edn., Englewood Cliffs: Prentice Hall.

Hodder, J. E. and Riggs, H. E. 1985. Pitfalls in evaluating risky projects. *Harvard Business Review,* 85: 128–35.

Hodder, J. F. 1984. Evaluation of manufacturing investments: A comparison of US and Japanese practices. *Technical Report 84–8,* Department of Industrial Engineering and Engineering Management, Stanford University.

Hoffman, A. J. 1997. *From Heresy to Dogma.* San Francisco: New Lexington Press.

House, R. J. 1971. A path-goal theory of leader effectiveness. *Administrative Science Quarterly,* September: 321–38.

1976. A theory of charismatic leadership. In J. G. Hunt and L. L. Larson (eds.), *Leadership: The Cutting Edge.* Carbondale, IL: Southern Illinois University Press, pp. 189–207.

Hout, T., Porter, M. and Rudden, E. 1982. How global companies win out. *Harvard Business Review,* 60: 98–108.

Ikenberry, G. J. 1993. Creating yesterday's new world order. In J. Goldstein and R.

O. Keohane (eds.), *Ideas and Foreign Policy*. Ithaca, NY: Cornell University Press, pp. 57–86.

International Organization for Standardization. 1995a. *ISO/DIS 14001: Environmental management systems–general guidelines on principles, systems, and supporting techniques*. Draft. Geneva: International Organization for Standardization.

1995b. *ISO/DIS 14004: Environmental management systems–general guidelines on principles, systems, and supporting techniques*. Draft. Geneva: International Organization for Standardization.

Jaffe, B. A., Peterson, S. R., Portney, P. R., and Stavins, R. N. 1995. Environmental regulation and the competitiveness of US manufacturing: What does the evidence tell us? *Journal of Economic Literature*, 33: 132–63.

Jensen, M. 1983. Organization theory and methodology. *Accounting Review*, 58: 319–39.

Jensen, M. and Meckling, W. 1976. Theory of firm: Managerial behavior, agency cost, and ownership structure. *Journal of Financial Economics*, 3: 305–60.

Johnson, P. L. 1994. *ISO 9000: Meeting the New International Standards*. McGraw-Hill.

Jones, T.M. 1995. Instrumental stakeholder theory: A synthesis of ethics and economics. *Academy of Management Review*, 20: 404–37.

Katz, D. and Kahn, R. L. 1966. *The Social Psychology of Organizations*. New York: Wiley.

Katz, M. L. and Shapiro, C. 1983. *Network Externalities, Competition, and Compatibility*. Princeton, NJ: Woodrow Wilson School, Princeton University.

Katz, M. L. and Shapiro, C. 1985. *Technology Adoption in the Presence of Network Externalities*. Princeton, NJ: Woodrow Wilson School, Princeton University.

Katz, R. 1974. Skills of an effective administrator. *Harvard Business Review*, September– October: 90–101.

Katzmann, R. 1980. *Regulatory Bureaucracy: The Federal Trade Commission and the Antitrust Policy*. Cambridge, MA: MIT Press.

Kulatilaka, N. and Marcus, A. J. 1992. Project valuation under uncertainty: When does DCF fail? *Journal of Applied Corporate Finance* 5: 92–100.

Keohane, R. O. 1988. International institutions: Two approaches. *International Studies Quarterly*, 32: 379–96.

Kester, W. C. 1996. American and Japanese corporate governance: Convergence to best practice. In S. Berger and R. Dore (eds.), *National Diversity and Global Capitalism*. Ithaca, NY: Cornell University Press, pp.107–37.

Khademian, A. M. 1992. *The SEC and Capital Market Regulation: The Politics of Expertise*. Pittsburgh: University of Pittsburgh Press.

Khanna, M. and Damon, L.A. 1999. EPA's voluntary 33/50 program: Impact on toxic releases and economic performances of firms. *Journal of Environmental Economics and Management*, 37: 1–25.

King, G., Keohane, R. O., and Verba, S. 1994. *Designing Social Inquiry*. Princeton, NJ: Princeton University Press.

Kiser L. L. and Ostrom, E. 1982. Three worlds of collective action: A metatheoretical synthesis of institutional approaches. In V. Ostrom (ed.), *Strategies for Political Inquiry*. Beverly Hills, CA: Sage, pp. 179–222.

Kissinger, H. A. 1964. *A World Restored.* New York: Grosset and Dunlap.

Knight, J. 1992. *Institutions and Social Conflict*. Cambridge University Press.

Kolko, G. 1963. *Railroads and Regulation, 1877–1916.* Princeton, NJ: Princeton University Press.

Kotter, J. and Heskett, J. 1992. *Corporate Culture and Performance*. New York: Free Press.

Kreps, D. M. 1990. Corporate culture and economic theory. In J. Alt and K. Shepsle (eds.), *Perspectives on Positive Political Economy*. Cambridge University Press, pp. 90–143.

Lee, D. and Misiolek, W.S. 1986. Substituting pollution taxation for general taxation: Some implications for efficiency in pollution taxation. *Journal of Environmental Economics and Management,* 13: 338–47.

Leibenstein, H. 1966. Allocative efficiency and X efficiency. *The American Economic Review,* 56: 392–415.

Leonard, J. H. 1988. *Pollution and the Struggle for the World Product: Multinational Corporations, the Environment, and International Comparative Advantage.* Cambridge University Press.

Libecap, G. D. 1989. *Contracting for Property Rights*. Cambridge University Press.

Lilly. 1992a. *A Perspective on Pricing*. Indianapolis, IN: Eli Lilly and Company.

1992b. *Environmental Annual Report*. Indianapolis, IN: Eli Lilly and Company.

1993. *Environmental Annual Report*. Indianapolis, IN: Eli Lilly and Company.

1994a. *Annual Report*. Indianapolis, IN: Eli Lilly and Company.

1994b. *Environmental Annual Report*. Indianapolis, IN: Eli Lilly and Company.

1994c. *Responsible Care: A Public Commitment*. Indianapolis, IN: Eli Lilly and Company.

1995a. *Environmental, Health, and Safety Report.* Indianapolis, IN: Eli Lilly and Company.

1995b. *Information 1995*. Indianapolis, IN: Eli Lilly and Company.

1995c. *Environmental Annual Report*. Indianapolis, IN: Eli Lilly and Company.

1996a. *The Three Loop Operations Design Model for Environmental, Health, and Safety Programs.* Indianapolis, IN: Eli Lilly and Company.

1996b. *Environmental Policies and Guidelines.* Indianapolis, IN: Eli Lilly and Company.

1997a. http://www.lilly.com/financial/95-annual/html/hilites.html; 03/02/97.

1997b. http://www.lilly.com/environment/96update/96gr9_1.htm; 05/02/97.

Lipman-Blumen, J. 1996. *The Connective Edge: Leading in an Interdependent World.* Jossey-Bass.

Lowi, T. J. 1969. *The End of Liberalism: The Second Republic of the United States*. 2nd edn, New York: Norton.

Luthans, F. 1995. *Organizational Behavior.* Singapore: McGraw-Hill, Inc.

Majumdar, R. C., Raychaudhuri, H. C., and Datta, K. 1958. *An Advanced History of India*. 2nd edn, London: Macmillan.

Maloney, M. T. and McCormick, R. E. 1982. A positive theory of environmental quality regulation. *Journal of Law and Economics,* 25: 99–123.

Manne, H. 1965. Mergers and the market for corporate control. *Journal of Political Economy,* 73: 110–120.

March, J. and Simon, H. A. 1958. *Organizations*. New York: John Wiley.

March, J. G. and Olsen, J. P. 1989. *Rediscovering Institutions: The Organizational Basis of Politics*. New York: Free Press.

Marcus, A. A. 1984. *The Adversary Economy*. Westport, CO: Quorum Books.

Marglin, S. 1974. What bosses do? The origins and functions of hierarchy in capitalist production. *The Review of Radical Political Economy*, 6: 33–60.

Marris, R. 1964. *The Economic Theory of Managerial Capitalism*. London: MacMillan.

Maslow, A. H. 1943. A theory of human motivation. *Psychological Review*, July: 370–96.

Meyer, J. and Scott, W.R. (eds.) 1992. *Organizational Environments: Ritual and Rationality*. Thousand Oaks, CA: Sage.

Miller, G. J. 1992. *Managerial Dilemmas: Political Economy of Hierarchy*. Cambridge University Press.

Mitchell, R. K., Agle, B.R., and Wood, D.J. 1997. Towards a theory of stakeholder identification and salience: Defining the principle who and what really counts. *Academy of Management Review*, 22: 853–86.

Morandi, L. and Pascal, S. 1995. *Environmental Audits: Incentives to Comply with to Avoid Regulation*? February, Washington, DC : National Conference of State Legislatures.

Morgenthau, H. J. 1978. *Politics Among Nations: The Struggle for Power and Peace*. New York: Knopf.

Mueller, D. C. 1989. *Public Choice II*. Cambridge University Press.

National Academy of Public Administration (NAPA). 1986. *A Strategy for Implementing Federal Regulations of Underground Storage Tanks*. Washington, DC : National Academy of Public Administration.

Nehrt, C. 1998. Maintainability of first-mover advantages when environmental regulations differ between countries. *Academy of Management Review*, 23: 77–97.

Nelson, R. R. and Winter, S. G. 1982. *An Evolutionary Theory of Economic Change*. Cambridge, MA: Harvard University Press.

Newton, T. and Harte, G. 1996. Green business: Technicist kitsch? *Journal of Management Studies*, 34: 75–98.

New York Times. 1996a. Many states give polluting firms new protection. April 17: A1, A16.

1996b. December 26: A26.

North, D. C. 1990. *Institutions, Institutional Change, and Economic Performance*. Cambridge University Press.

Northouse, P.G. 1997. *Leadership: Theory and Practice*. Thousand Oaks, CA: Sage.

NSF International. 1996. *Environmental Management System Demonstration Project: Final Report*. Ann Arbor, MI: NSF International.

Oates, W. E., Portney, P. R., and McGartland, A. M.1989. The net benefits of environmental regulations. *American Economic Review*, 79: 1233–42.

Oblak, D. J. and Helm, R. J. 1980. Survey and analysis of capital budgeting methods used by multinationals. *Financial Management*, 2: 37–40.

Oliver, C. 1991. Strategic responses to institutional processes. *Academy of Management Review*, 16: 145–79.

Olson, M. 1965. *The Logic of Collective Action–Public Goods and the Theory of Groups*. Cambridge, MA: Harvard University Press.

Osborne, D. and Gaebler, T. 1992. *Reinventing Government: How the Entrepreneurial Spirit is Transforming the Public Sector.* Cambridge, MA: Addison-Wesley.

Ostrom, E. 1986. An agenda in the study of institutions. *Public Choice,* 48: 3–25.

1990. *Governing the Commons: Evolution of Institutions for Collective Action.* Cambridge University Press.

1991. Rational-choice theory and institutional analysis: Towards complementarity. *American Political Science Review,* 85: 237–43.

1996. Institutional rational choice: An assessment of the IAD framework. Working Paper W96–16, Bloomington, IN: Workshop in Political Theory and Policy Analysis.

Ostrom, E., Gardner, R., and Walker, J. 1994. *Rules, Games, and Common-Pool Resources*. Ann Arbor, MI: University of Michigan Press.

Ostrom, V. and Ostrom, E. 1977. Public goods and public choice. In E. S. Savas (ed.), *Alternatives for Delivering Public Services*. Boulder, CO: Westview, pp. 7–49.

Palmer K., Oates, W., and Portney, P. R. 1995. Tightening environmental standards: The benefit–cost or the no–cost paradigm? *Journal of Economic Perspectives,* 9: 119–32.

Pearson, C. S. 1985. *Down to Business: Multinational Corporations, the Environment, and Development*. Washington, DC: World Resource Institute.

Penrose, E. T. 1959. *The Theory of the Growth of the Firm*. London: Basil Blackwell.

Perrow, C. 1979. *Complex Organizations: A Critical Essay*. Glenview, IL: Scott, Foresman and Company.

Peteraf, M. 1993. The cornerstone of competitive advantage: A resource-based view. *Strategic Management Journal,* 14: 179–92.

Pfeffer, J. 1981. *Power in Organizations*. Marshfield, MA: Pitman.

Pfeffer, J. and Salanick. 1978. *The External Control of Organization*. New York: Harper and Row.

Pigou, A.C. 1960. *The Economics of Welfare*. London: Macmillan.

Platt, J. 1973. Social traps. *American Psychologist,* 26: 641–51.

Porter, M. E. 1991. America's green strategy. *Scientific American,* April: 168.

Porter, M. E. and van der Linde, C. 1995. Toward a new conception of the environment-competitiveness relationship. *Journal of Economic Perspectives,* 9: 97–118.

Prahalad, C.K. and Hamel, G. 1990. The core competence of the corporation. *Harvard Business Review,* 68: 79–91.

Prakash, A. 1999a. A new-institutionalist perspective on ISO 14000 and Responsible Care. *Business Strategy and the Environment*. Forthcoming.

1999b. Why firms adopt beyond-compliance policies. Mimeo.

Prakash, A. Krutilla, K., and Karamanos, P. 1996. Multinational corporations and international environmental policy. *Business and the Contemporary World,* 8: 119–44.

Preston, L.E. 1975. Corporation and society: The search for a paradigm. *Journal of Economic Literature,* 13: 434–53.

Preston, L.E. and Post, J.E. 1975. *Private Management and Public Policy: The Principle of Public Responsibility*. Englewood Cliffs, NJ: Prentice-Hall.

Preston, L.E. and Sapienza, H. J. 1990. Stakeholder management and corporate performance. *Journal of Behavioral Economics,* 19: 361–75.

Puri, S. C. 1996. *Stepping Up to ISO 14000: Integrating Environmental Quality with ISO 9000 and TQM*. Portland, OR: Productivity Press.

Putterman, L. 1984.On some recent explanations of why capital hires labor. *Economic Inquiry,* 22: 171–87.

Reitz, H. J. 1987. *Behavior in Organizations*. Homewood, IL.: Irwin.

Roberts, P. W. and Greenwood, R. 1997. Integrating transaction costs and institutional theories. *Academy of Management Review,* 22: 346–73.

Ross, S.A. 1974. The economic theory of agency: The principal problem. *American Economic Review,* 62: 134–9.

Rugman, A.R. and Verbeke, A. 2000. Environmental regulations and global strategies of multinational corporations. In A. Prakash and J. A. Hart(eds.), *Coping with Globalization*. London and New York: Routledge.

Russo, M.V. and Fouts, P.A. 1997. A resource-based perspective on corporate environmental performance and profitability. *Academy of Management Journal,* 40: 534–59

Saloner, Garth and Shepard, Andrea. 1991. *Adoption of Technologies with Network Effects: An Empirical Examination of the Adoption of Automated Teller Machines*, Stanford, CA: Graduate School of Business.

Salop, S.C. and Scheffman, D.T. 1983. Raising rivals' costs. *American Economic Association, Papers and Proceedings,* 73: 267–71.

Samuelson, P. A. 1947. *Foundations of Economic Analysis*. Cambridge, MA: Harvard University Press.

Sandborg, V. 1996. cited in "Baxter looking at ISO 14000 but favors own system." *Integrated Management System Update,* June: 23.

Sandholtz, W. 1999. Globalization and the evolution of rules. In A. Prakash and J. A. Hart (eds.), *Globalization and Governance*, London and New York: Routledge.

Sarkar, S. 1989. *Modern India 1885–1947*. New York: St. Martin's Press.

Sarokin, D. 1999. Email conversation. February 11.

Schall, L.D., Sundem, G. L., and Geijsbeek, W. R. 1978. Survey and analysis of capital budgeting methods. *Journal of Finance,* 33: 281–87.

Schelling, T. 1978. *Micromotives and Macrobehavior*. New York: Norton.

Schmidheiny, S. in collaboration with the Business Council for Sustainable Development. 1992. *Changing Course: A Global Business Perspective on Environment and Development*. Cambridge, MA: MIT Press.

Schmidheiny, S. and Zorraquin, F.J.L. (in collaboration with the Business Council for Sustainable Development). 1996. *Financing Change*. Cambridge, MA: MIT Press.

Schneider, M. and Teske, P. with Mintrom, M. 1995. *Public Entrepreneur: Agents for Change in American Government*. Princeton, NJ: Princeton University Press.

Scott, W. R. 1987. The adolescence of institutional theory. *Administrative Science Quarterly,* 32: 493–511.

Selznick, P. 1957. *Leadership in Administration*. Evanston, IL: Row, Peterson, and Company.

Senge, P. M. 1994. *The Fifth Discipline: The Area and Practice of the Learning Organization*. New York: Doubleday/Currency.

Shrivastava, P. 1995. The role of corporation in achieving ecological sustainability. *Academy of Management Review,* 20: 936–60.

Simon, H. A. 1957. *Models of Man*. New York: John Wiley.

Simmons, P. and Wynne, B. 1993. Responsible Care: Trust, credibility, and environmental management. In K. Fischer and J. Schot (eds.), *Environmental Strategies for Industry.* Washington, DC: Island Press, pp. 201–26.

Smith, A. 1970. *The Wealth of Nations: Books I–III*. London: Pelican Books.

Snidal, D. 1995. The politics of scope: Endogenous actors, heterogeneity, and institutions. In R. O. Keohane and E. Ostrom (eds.), *Local Commons and Global Interdependence*. London: Sage, pp. 47–70.

Stigler, G. 1971. The economic theory of regulation. *Bell Journal of Economics,* 2: 3–21.

Thompson, A. R. 1973. *Economics of Firm*. Engelwood Cliffs, NJ: Prentice Hall.

Tiebout, C. M. 1956. A pure theory of local expenditure. *Journal of Political Economy,* 64: 416–24.

Tietenberg, T. H. 1992. *Environmental and Natural Resource Economics*. New York: Harper Collins.

Tolbert, P. S. 1985. Resource dependence and institutional environment: Sources of administrative structures in institutions of higher education. *Administrative Science Quarterly,* 30: 1–13.

United Nations Conference on Trade and Development/UNCTAD. 1993. *Environmental Management in Transnational Corporations: Report on the Benchmark Corporate Environmental Survey*. New York and Geneva: United Nations.

1997, *World Investment Report,* 1997. New York and Geneva: United Nations.

US Environmental Protection Agency (USEPA). 1987. Underground storage tanks: Technical requirement. 52 *Federal Register* 12662, 12664, April 7.

1988a Underground storage tanks containing hazardous substances: Financial responsibility requirement, advance notice of proposed rulemaking. 53 *Federal Register* 3817, 3819, February 9.

1988b. Technical requirement: Final rule. 53 *Federal Register* 37082, 37088, September 23.

Vangeloff, J. 1992. Above ground small volume solvent storage. In *Above Ground storage tanks,* Proceedings of the Second International Symposium on Above Ground Tanks, January 14–16, 1992, Houston, Texas, pp.5/1–5/8.

Vogel, D. 1986. *National Styles of Regulation*. Ithaca, NY: Cornell University Press.

1995. *Trading Up.* Cambridge, MA: Harvard University Press.

Walters, I. 1973. The pollution content of American trade. *Western Economic Journal,* 11: 61–70.

Walley, N. and Whitehead, B. 1994. It's not easy being green. *Harvard Business Review,* May–June: 46–52.

Wartick, S.L. and Cochran, P.L. 1985. The evolution of the corporate social performance model. *Academy of Management Review* 4: 758–69.

Weber, M. 1947. *The Theory of Social and Economic Organization*. New York: Free Press.

Weick, K. E. 1995. *Sensemaking in Organizations*. London: Sage.

Weimer, D. L. (ed.) 1997. *The Political Economy of Property Rights*. Cambridge University Press.

Weimer, D. L. and Vining, A.R. 1992. *Policy Analysis: Concepts and Practice*. Englewood Cliffs, NJ: Prentice Hall.

Wendt, A. 1992. Anarchy is what the states make of it: The social construction of power politics. *International Organization*, 46: 391–425.

Wicks, M. E. 1980. *A Comparative Analysis of Foreign Investment Evaluations Practices of US Based Multinational Corporations*. New York: McKinsey and Company.

Wilkins, J. R. 1996 Lilly fulfills pledge to reduce emission, A note prepared for submission to the EPA. July 2, 1996. Indianapolis, IN: Eli Lilly and Company.

Williamson, O. E. 1964. *The Economics of Discretionary Behavior*. Englewood Cliffs, NJ: Prentice Hall.

1975 *Market and Hierarchies: Analysis and Antitrust Implications*. New York: Free Press.

1985. *Economic Institutions of Capitalism*. New York: Free Press.

Wiseman, J. 1957. The theory of public utility price – An empty box. *Oxford Economic Papers*, 9: 56–74.

Wood, D.J. 1991. Corporate social performance revisited. *Academy of Management Review*, 16: 691–718.

Wood, D.J. and Jones, R.E. 1995. Stakeholder mismatching: A theoretical problem in empirical research on corporate social performance. *The International Journal of Organizational Analysis*, 3: 229–67.

World Bank. 1992. *World Development Report*. Washington, DC: World Bank; Oxford: Oxford University Press.

World Commission on Environment and Development. 1987. *Our Common Future*. Oxford: Oxford University Press.

Yukul, G. A. 1981. *Leadership in Organization*. Englewood Cliffs, NJ: Prentice Hall.

Zinn, H. 1995. *A People's History of the United States: 1492 to Present*. New York: Harper Perennial.

Zucker, L. G. (ed.) 1988. *Institutional Patterns and Organization*. New York: Ballinger.

Name index

Subject index